Excel Manual

for

Moore and McCabe's

Introduction to the Practice
of Statistics

Fourth Edition

Fred M. Hoppe
McMaster University

W. H. Freeman and Company
New York

Printed in the United States of America

ISBN: 0-7167-4915-7

Second printing, 2002

Contents

Preface

In recent years there has been a paradigm shift in the teaching of statistics away from a mathematical presentation and toward an emphasis on the exploratory and data analytic nature of the subject, often with reference to substantive fields.

A major theme in this new statistical symphony is that results of any analysis be supplemented with a graph. Another theme is that students who complete a first course should be able to practically, not merely theoretically, do a statistical analysis. Given a data set and a set of questions to be addressed, students should know how to input the data using the appropriate tool (from the list of tools they have been taught), examine the data graphically for evidence that confirms or dispels their assumptions, draw justifiable conclusions, and finally export the results for presentation in report form.

Central then to the success of such an approach to teaching statistics is reliable and easy-to-learn software. Many excellent software packages do exist but contain encyclopedic user manuals unsuitable for a first course at the undergraduate level. Even massive packages are not suited for all purposes. For instance, although SAS can do logistic regression, the best tool for the task is S-plus.

For a number of years at McMaster University I have taught a statistics course taken by first-year science students who pine for a career in the life sciences. Statistics is often a prerequisite for upper-level courses. Computing is a central and tested aspect of this course because students are expected to carry out statistical analyses in laboratory courses involving real data. To this end, small groups of students meet in computer labs to learn at least one statistics package.

The software we have traditionally used in first-year courses at McMaster is Minitab (in higher years, students are exposed to SAS and S-plus, as well as Matlab) because it is relatively easy to learn and McMaster has a site license for Minitab within its computer network. Unfortunately, the cost of legitimate individual copies of Minitab, even the student version, makes purchase of the software prohibitive for most students. This means that the only time many of them have for using Minitab is in the lab.

I have noticed that the reports submitted by some students, especially for assignments that require graphical output, contain more sophisticated graphs than are available with Minitab. I have learned that many choose to work in Excel through Microsoft Office instead of Minitab.

Excel is a spreadsheet program, a tool for organizing data contained in columns and rows. Operations on data mimic those described by mathematical functions, and the formulas required for data analysis can therefore be expressed as spreadsheet operations. Although originally developed as a business application to display numbers in a tabular format and to automatically recalculate values in response to changes in numbers in the table, current spreadsheets enjoy built-in functions and display capabilities that can be used for statistical analysis.

Spreadsheets are therefore an alternative to specialized statistical software. Because Excel is part of an integrated word processing/graphics/database package, data can be easily input from other applications and exported into report form. For instance, I regularly download stock prices into Excel from the Internet to plot the course of investments, and my students enjoy this particular application as an incentive for learning statistics.

For the last four years I have used both *Introduction to the Practice of Statistics* by David Moore and George McCabe and *The Basic Practice of Statistics* by David Moore. Previously, I had used another book by David Moore called *Statistics, Concepts and Controversies* when I was at the University of Michigan, Ann Arbor. These books are leading expositions of the new approach to the teaching of statistics.

So when Christopher Spavins, the W. H. Freeman sales representative at McMaster, with whom I've had many interesting conversations on the teaching of statistics, asked if I was interested in developing an Excel manual to accompany the third edition of *Introduction to the Practice of Statistics (IPS)*, I agreed. This guide, produced over a span of two months, is the result.

Some key features of this guide:

- Introductory chapter on Excel for those with no prior knowledge of Excel.
- Written for Excel 97/98 with an Appendix for Excel 5/95. Explanations are given in parallel for Excel 97/98 and Excel 5/95 when the interface differs.
- Figures from Excel 98 for Macintosh, Excel 97 for Windows, and Excel 5 for Macintosh showing differences and similarities in the user interface.
- Presentation follows *IPS* with fully worked and cross-referenced examples and exercises from *IPS*.
- Detailed exposition of the ChartWizard for graphical displays.
- Extensive use of Named Ranges to make formulas more transparent.
- Nearly 200 figures accompanied by step-by-step descriptions.
- Detailed use of simulation to explain randomness by simulating the Central Limit Theorem and the Law of Large Numbers. General distributions are simulated from the uniform (0,1) distribution by using critical values obtained in Excel to find the inverse probability transformation.
- Construction of a normal table and graphs of the normal and the Student t densities by a general method applicable to other densities.
- Development of weighted least-squares for logistic regression.

- Templates provided for one-sample procedures, chi-square tests, and nonparametric statistics.
- Development of side-by-side boxplots to supplement Excel.

I believe that Excel can satisfy all of a student's needs in a course based on a book such as *IPS*. In fact, every technique discussed in *IPS* can be developed within Excel. Students who learn statistics using Excel will find it extremely valuable in future courses (a point brought home by one of my third-year students who informed me that she used Excel in her physics course for plotting data because she "didn't feel like doing it by hand").

In some of my own consulting I have found researchers, self-taught in statistics, using Excel. Much of experimental research is exploratory in nature and Excel's excellent graphical capabilities and easy-to-use interface make it the statistics tool of choice for many in the life sciences, engineering, and business. Finally, given that Excel is produced by Microsoft, one can predict without fear a long, useful, and upgradeable life for Excel. All these considerations lead me to believe that Excel will grow in popularity in the teaching of statistics.

I will be putting up a Web site at http://www.mathematics.net/ where updates, macros, and other useful information will appear. If you have any material you think may be useful for this site, contact me. It is also my intention to install a bulletin board to facilitate communication by faculty and students who use this manual with *IPS*.

In addition to Chris Spavins, I thank Patrick Farace, statistics editor at W. H. Freeman, for his assistance in bringing this book to market and for showing me life in the fast lane of publishing, and I am grateful to Erica Seifert and Jodi Isman for their editorial and copyediting skills. The book was typeset in LaTeX by Debbie Iscoe, who, as with my earlier book, did a superb job with both care and good cheer. Brian Golding kindly showed me how to include eps files as figures.

Finally, thanks to my terrific wife, Marla, and super kids, Daniel and Tamara, for putting up with a kitchen table that became the extension of my desk and for lovingly and unquestioningly giving me the time (though not necessarily quiet time) to complete this project.

The Fourth Edition

This edition of the manual has been updated for Excel 2000 (Windows) and Excel 2001 (Macintosh) with all images redone. Many new examples have been worked out, all of them still referencing either examples or exercises in the text in order to facilitate coordination of the manual and text. Misprints in the previous edition were also corrected.

This manual may still be safely adopted by users of Excel 97 (Windows) or Excel 98 (Macintosh) as Excel 2000/2001 did not make substantial changes from Excel 97/98. There were improvements in the user interfaces and in some features such as the PivotTable, but the main changes dealt with VBA programming and integration with the Internet, for instance in allowing workbooks to be saved as web pages (html files). All of the instructions provided were checked for compatibility with Excel 97/98. As with the previous edition, any differences between Windows and Macintosh are made explicit and also for Excel 5/95 there are separate instructions to cover some deviations from Excel 97/98/2000/2001. Users of the recent version of Excel 2002, part of the Office XP suite, should also be fine because again only minor changes were implemented in Excel 2002.

Adopters of this manual have requested that macros be supplied to supplement some of the features lacking in Excel. With this edition we have begun to do this and make them available at the Freeman Web site (http://www.freeman.com/ips/). These macros will be updated from time to time.

I would like to thank Mark Santee, Brian Donnellan, and Danielle Swearengin, all editors at W. H. Freeman with whom I have worked, and my family for their continued encouragement.

FRED M. HOPPE
DUNDAS, ONTARIO
JUNE 17, 2002
E-mail: *hoppe@mcmaster.ca*

Introduction

This book is a supplement to *Introduction to the Practice of Statistics*, Fourth Edition, by David S. Moore and George P. McCabe, referred to as *IPS*. Its purpose is to show how to use Excel in performing the common statistical procedures in *IPS*.

I.1   What Is Excel?

Microsoft Excel is a spreadsheet application whose capabilities include graphics and database applications. A spreadsheet is a tool for organizing data. Originally developed as a business application for displaying numbers in a table, numbers that were linked by formulas and updated whenever any part of the data in the spreadsheet changed, Excel now has built-in functions, tools, and graphical features that allow it to be used for sophisticated statistical analyses.

Windows or Macintosh?

It doesn't matter which you use. This book is designed equally for Macintosh or Windows operating systems. The Macintosh and Windows versions of Excel function essentially the same way, with a few slight differences in the file, print, and command shortcuts. These are due mainly to the absence of a right mouse button for the Macintosh. However, the right button action can be duplicated with a keystroke, and I have described both actions where they differ. Both Macintosh and Windows users will find this book useful.

Nearly all figures shown in this book have been generated using Excel 2001 on a Macintosh G4 running Mac OS 10.1.4. Corresponding figures from the Windows version are only cosmetically different, for instance in Fig 1.1 in Chapter 1 showing the opening screen of the ChartWizard. Students should feel familiar with the look of the Excel interface no matter what the platform.

Which Version of Excel Should I Use?

Naming conventions are slightly confusing because of a plethora of patches, bug fixes, interim releases, and so on. In the Windows environment the main versions used are Excel 5.0, 5.0c, Excel 7.0, 7.0a (for Windows 95), Excel 8.0 (also called

Excel 97), Excel 2000, and the recent Excel 2002. For Macintosh, they are Excel 5.0 and 5.0a, Excel 98, and Excel 2001.

The major change in the development occurred with Excel 5.0. The statistical tools in Excel 5.0 and Excel 7.0 for Windows and Excel 5.0 for Macintosh function in virtually the same way. These are referred to collectively in this book as Excel 5/95. Likewise, the Windows versions Excel 97, Excel 2000 and the Macintosh versions Excel 98, Excel 2001 function similarly and are referred to as Excel 97/98/2000/2001.

Excel 97/98 removed some bugs in the Data Analysis ToolPak (but introduced others), improved the interface to the ChartWizard, and replaced the Function Wizard with the Formula Palette. Help was greatly expanded with the introduction of the animated Office Assistant. Minor cosmetic changes were made in the placement of components within some dialog boxes. Although changes were made in the VBA interface, few substantive changes occurred that might cause differences in execution or statistical capabilities between Excel 5/95 and Excel 97/98. Both Excel 2000 and Excel 2002 introduced only incremental improvements, mostly in the user interface rather than programming.

This book is based on Excel 2000/2001 but all the instructions however have been tested on Excel 97/98. Moreover, given the existing base of Excel 5/95 users I have tried to make this book equally accessible to them without having two separate editions. Surprisingly, this has not been too difficult. The main differences needing instructional care are the description of the ChartWizard (reduced to four steps in Excel 97/98 from five steps in Excel 5/95) and the implementation of formulas with the Formula Palette (Excel 97/98) instead of the Function Wizard (Excel 5/95). In addition, there are related differences in some pull-down options from the Menu Bar. On the other hand, the Data Analysis ToolPak is virtually the same in each version. Nearly all differences are in Chapters 1 and 2. While earlier editions of this manual included a detailed Appendix covering these two chapters for Excel 5/95, it has been removed since a user of Excel 5/95 can easily make modifications based on that older interface. In the remaining chapters, whenever there is a technique whose implementation varies between the two versions, I have given separate steps. Usually only a few lines of text (four or fewer) suffice, underscoring the similarity in the versions in using Excel for statistics. It is hoped that this dual approach will make this manual truly equally useful for all versions from Excel 5/95 to Excel 2000/2001.

Do I Need Prior Familiarity with Excel?

The short answer is no. This book is completely self-contained. This introductory chapter contains a summary and description of Excel that should provide enough detail to enable a student to get started quickly in using Excel for statistical calculations. The subsequent chapters give step-by-step details on producing and embellishing graphs, using functions, and invoking the **Analysis Toolbox**. For users without prior exposure to Excel, this book may serve as a gentle exposure to spreadsheets and a starting point for further exploration of their features.

I.2   The Excel Workbook

When you first open Excel, a new file is displayed on your screen. Fig I.1 shows an Excel 2001 Macintosh opening workbook.

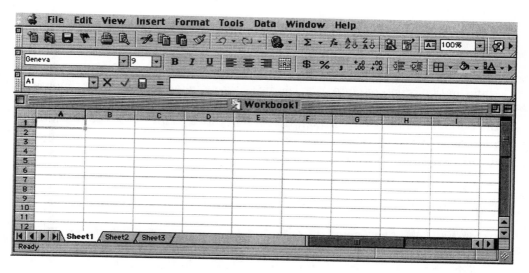

Figure I.1: Excel 2001 Macintosh Workbook

A workbook consists of various sheets in which information is displayed, usually related information such as data, charts, or macros. Sheets may be named and their names will appear as tabs at the bottom of the workbook. Sheets may be selected by clicking on their tabs and may be moved within or between workbooks. To keep the presentation simple in this book we have chosen to use]]] one sheet per workbook in each of our examples.

A sheet is an array of cells organized in rows and columns. The rows are numbered from 1 to 65,536 (up from 16,384 in Excel 5/95) while the columns are described alphabetically, as follows,

$$A, B, C, \ldots, X, Y, Z, AA, AB, AC, \ldots, IV$$

for a total of 256 columns.

Each cell is identified by the column and row that intersect at its location. For instance, the selected cell in Fig I.1 has address (or cell reference) A1. When referring to cells in other sheets, we need to also provide the sheet name. Thus Sheet2!D9 refers to cell D9 on Sheet 2.

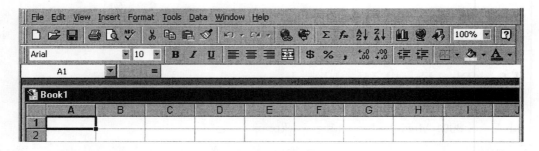

Figure I.2: Excel 97 Windows

Information is entered into each cell by selecting an address (use your mouse or arrow keys to navigate among the cells) and entering information, either directly in the cell or else in the **Formula Bar** text entry area. Three types of information can be entered: labels, values, and formulas. We will discuss entering information in detail later in this chapter.

I.3   Components of the Workbook

Look at Fig I.1. Then compare with Fig I.2 showing an Excel 97 (Windows) workbook and Fig I.3, an Excel 5 Macintosh workbook.

There are two main components in a workbook: the document window and the application window. Information is entered in the document window identified by the row and column labels. Above the document window are all the applications, functions, tools, and formatting features that Excel provides. There are so many commands in a workbook that an efficient system is needed to access them. This is achieved either by pull-down menus invoked from the Menu Bar or from equivalent icons on the toolbars.

The application window is thus the control center from which the user gives instruction to Excel to operate on the data in the document window below it. We

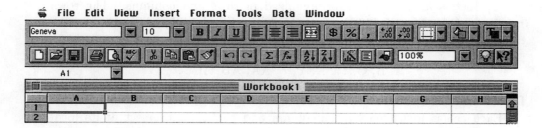

Figure I.3: Excel 5 Macintosh

| File | Edit | View | Insert | Format | Tools | Data | Window | Help |

Figure I.4: Menu Bar

Table I.1: Menu Bar Pull-Down Options

Menu Bar	Pull-Down Options
File	Open, close, save, print, exit
Edit	Copy, cut, paste, delete, find, etc. (basic editing)
View	Controls which components of workbook are displayed on screen, size, etc.
Insert	Insert rows, columns, sheets, charts, text, etc. into workbook
Format	Format cells, rows, columns
Tools	Access spelling macros, data analysis toolpak (will be used throughout to access the statistical features of Excel)
Data	Database functions such as sorting, filtering
Window	Organize and display open workbooks
Help	Online help (also available on the **Standard Toolbar**)

examine the main components of the application window in detail.

Menu Bar

The **Menu Bar** (Fig I.4) appears at the top of the screen. It provides access to all Excel commands: **File, Edit, View, Insert, Format, Tools, Data, Window, Help**. Each word in the Menu Bar opens a pull-down menu of options familiar to users of any window-based application (there are also keyboard equivalents). As the name implies, this is the main component of the control center and will be elaborated on in various examples throughout this book. Table I.1 summarizes some of the options available.

Toolbars

When Excel is opened, two strips of icons appear below the Menu Bar: the **Standard Toolbar** and the **Formatting Toolbar**. Other toolbars are available by choosing **View − Toolbars** from the Menu Bar and making a selection from the choices available. Existing toolbars can be customized by adding or removing but-

Figure I.5: Excel 2001 Standard Toolbar

tons, and new ones can be created. For the purpose of this book you will not need to make such customizations.

Standard Toolbar

The (default) **Standard Toolbar** (Fig I.5) provides buttons to ease your access to basic workbook tasks. Included are buttons for the following tasks:

- Start a new workbook
- Open existing workbook
- Save open workbook
- Print, print preview
- Check spelling
- Cut selection and store in clipboard for posting elsewhere
- Copy selected cells to clipboard, paste data from clipboard
- Copy format
- Undo last action, redo last action
- Insert hyperlink
- World Wide Web interface
- Autosum function (may also be entered directly into cell or **Formula Bar**)
- **Paste Function** (step-by-step dialog boxes to enter a function connected to the **Formula Palette**)
- Sort descending, sort ascending order
- **ChartWizard** (covered in detail in Chapter 1)
- **PivotTable Wizard**
- Drawing toolbar
- Zoom factor for display
- **Office Assistant**

As you pass over a button with your mouse pointer a small text label appears next to the button.

Formatting Toolbar

Fig I.6 shows the **Formatting Toolbar,** by which you can change the appearance of text and data. Features offered:

- Display and select font of selected cell

- Apply bold, italic, underline formatting
- Left, center, or right justify data
- Merge and center
- Apply currency style, percent style, etc.
- Increase or decrease decimal places
- Indent
- Add borders to selected sides of cell
- Change background color of cell, change color of text in cells

Figure I.6: Formatting Toolbar

Formula Bar

The **Formula Bar** is located just above the document window (Fig I.7). There are six areas in the Formula Bar (from left to right):

Figure I.7: Formula Bar

- **Name box.** Displays reference to active cell or function.
- **Defined name pull-down.** Lists defined names in workbook.
- **Cancel box.** Click on the red X to delete the contents of the active cell.
- **Enter box.** Click the green check mark to accept the formula bar entry.
- **Formula Palette (Excel 97/98 and Excel 2000/2001).** Constructs a function using dialog boxes to access Excel's built-in functions, or the function can be entered directly if you know the syntax. (Excel 5/95 has f_x in place of the equal (=) sign to activate the **Function Wizard**).
- **Text/Formula Entry area.** Enter and display the contents of the active cell.

The **Cancel** box and **Enter** box buttons appear only when a cell is being edited. Once the data have been entered, they disappear.

Title Bar

This is the name of your workbook. On a Mac the default is **Workbook1**, which appears just above the document window. With Windows the default name is **Book1**.

Document (Sheet) Window

A sheet in a workbook contains 256 columns by 65,536 rows. Use the mouse pointer or arrow keys to move from cell to cell. The pointer may change appearance depending on what actions are permitted. It might be an arrow, a blinking vertical cursor (I-beam), or an outline plus sign, for instance.

Sheet Tabs

A single workbook can have many sheets, the limit determined by the capacity of your computer; it is sometimes convenient to organize a workbook with multiple sheets, for instance sheets for data, analyses, bar graphs, or Visual Basic macros. Each sheet has a tab located at the bottom of the workbook (Fig I.8), and a sheet is activated when you click on its tab. Tab scrolling buttons allow you to navigate among the sheets. Clicking on a sheet tab on the bottom of the sheet activates it. Each workbook consists initially of three sheets labeled Sheet1, Sheet2, Sheet3 (16 initially in Excel 5). Sheets can be added, deleted, moved, and renamed to achieve a logical organization of data and analyses. To rename, move, delete, or copy a sheet, **right-click (Windows)** on a sheet tab or click and hold down the Control key on a **Macintosh**—we will refer to this as **Control-click**—and a pop-up menu appears from which you can select. A new sheet can also be inserted from the Menu Bar by choosing **Insert–Worksheet**. Note that a new worksheet is added to the left of the current or selected sheet. Sheets can also be deleted, copied, or edited from the Menu Bar using **Edit – Delete Sheet** or **Edit – Move or Copy Sheet. . . .**

   Another way to move a sheet is to grab it by clicking on it and holding the mouse button (left button for Windows). A small icon of a paper sheet will appear under the mouse pointer. As you move the mouse pointer, you will notice a small dark marker moving between the sheet tabs. This marker indicates where the sheet will be moved when you release the mouse button.

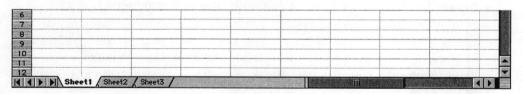

Figure I.8: Sheet Tabs

I.4   Entering and Modifying Information

When a workbook is first opened, cell A1 automatically becomes the active cell. Active cells are surrounded by a dark outline indicating that they are ready to receive data. Use the mouse (or arrow keys on the keyboard) to activate a different cell. Then enter the data and either click on the Enter box or press the enter (return) key.

Labels, Values, and Formulas

Three types of information can be entered into a cell: labels, values, and formulas. **Labels** are character strings such as words or phrases, typically used for headings or descriptions. They are not used in numerical calculations. **Values** are numbers such as 1.3, $1.75, π. **Formulas** are mathematical expressions that use the values or formulas in other cells to create new values or formulas. All formulas begin with an equal (=) sign and are entered directly by hand in the cell or in the text entry area of the **Formula Bar** or by the **Function Wizard**. As an example of how a formula operates, if the formula

$$= A1 + A2 + A3 + A4$$

is entered in cell A5 and if the contents of cells A1, A2, A3, A4 are 11, 12, 19, -6, respectively, then cell A5 will show the value 36 because what is displayed in the cell is the result of the computation, not the formula. The formula in the cell may be viewed in the entry area in the **Formula Bar** if the cursor is placed over the cell.

It is the existence of formulas that makes a spreadsheet such a powerful tool. A formula such as

$$= \text{SUM}(A1 : A4)$$

is the Excel equivalent of the mathematical expression

$$\sum_{i=1}^{4} A_i$$

and a complex mathematical expression can be rendered into an Excel workbook in a similar fashion.

Editing Information

There are several ways to edit information. If the data have not yet been entered after typing, then use the backspace or delete key or click on the red X to empty the contents of a cell. After the data have been entered, **activate** the cell by clicking on it. Then move the cursor to the text entry area of the Formula Bar where it turns into a vertical I-beam. Place the I-beam at the point you wish to edit and proceed to make changes.

Cell References, Ranges, and Named Ranges

A **cell reference** such as A10 is a **relative** reference. When a formula containing the reference A10 is copied to another location, the cell address in the new location is changed to reflect the position of the new cell. For instance, if the formula presented earlier

$$= A1 + A2 + A3 + A4$$

that appears in cell A5 is copied to cell D9, then it will become

$$= D5 + D6 + D7 + D8$$

to reflect that the formula sums the values of the four cells just above its location.

This relative addressing feature makes it relatively easy to repeat a formula across a row or column of a sheet, such as adding consecutive rows, by entering the formula once in one cell, then copying its contents to the other cells of interest.

If you need to retain the actual column or row label when copying a formula, then precede the label with a dollar sign ($). This is called an **absolute** cell reference. For instance, $A2 keeps the reference to column A but the reference to row 2 is relative, A$2 leaves A as a relative reference, but fixes the row at 2, $A$2 gives the entire cell (row and column labels) an absolute reference. We will use mixed ($A2 or A$2) references in Chapter 9 with the chi-square distribution.

A group of cells forming a rectangular block is called a **range** and is denoted by something like A2:B4, which includes all the cells {A2, A3, A4, B2, B3, B4}. **Named ranges** are names given to individual cells or ranges. The main advantage of named ranges is that they make formulas more meaningful and easy to remember. We will use named ranges repeatedly in this book. To illustrate, suppose we wish to refer to the range A2:A7 by the name "data." Enter the label "data" in cell A1. Then select the range A1:A7 by clicking on A1, holding the mouse button down, dragging to cell A7, and then releasing the mouse button. From the Menu Bar choose **Insert − Name − Create**, and check the box **Top Row** then click **OK** (Fig I.9).

Perhaps the quickest way to add a named range is to select the cell(s) to be named and then click on the **Name box**. This creates the name and associates it with the selected cells.

You can see which names are in your workbook by clicking on the defined name drop-down list arrow to the right of the **Name box**. Names are displayed alphabetically.

If you type a name from your workbook in the **Name** box and hit Enter then you will be transported to the first entry in the corresponding range, which will appear in the text/formula entry area of the Formula Bar ready for editing. Named ranges are both an aid in remembering and constructing formulas and also a convenient way to move around your workbook. They are used extensively in this book.

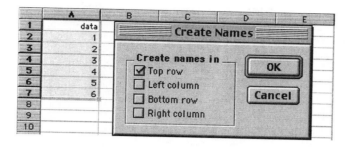

Figure I.9: Named Range

Copying Information

To activate a block of cells, place the cursor in the upper left cell of the block, click and drag to the lower right cell and release. Alternatively activate the upper left cell, then press the **Shift** key and click on the lower right cell. Noncontiguous cells or blocks can be selected by holding down the **Command** key (**Macintosh**) or the **Control** key (**Windows**) while selecting each successive cell or block.

To copy data from a cell or range, activate it, then from the Menu Bar choose **Edit – Copy**. Move the mouse cursor to the new location and from the Menu Bar choose **Edit – Paste**.

An alternative is to activate the range, then place the mouse cursor on the border of the selected range. It will appear as a pointer. Now press the **Control** key and move the cursor to the location for copying. Release the mouse button.

To move data to another location choose **Edit – Cut** from the Menu Bar and then **Edit – Paste** after you move the mouse cursor to the new location. Alternatively, move the mouse cursor to the border of the selected range. It turns into a pointer. "Grab" the border with the mouse pointer and move the cells to the desired location.

Use of the mouse for copying and cutting is a **drag and drop** operation familiar to users of Microsoft Word.

Shortcut Menu

If you activate a range and then **Control-click** (**Macintosh**) or **right-click** (**Windows**) on it, a **Shortcut Menu** will pop up next to the range allowing you access to some of the commands in the Menu Bar under **Edit**. This will provide options prior to copying or pasting.

Paste Special

A useful command from the **Shortcut Menu** is the **Paste Special**. This provides a dialog box (Fig I.10) giving a number of options prior to pasting. The two options

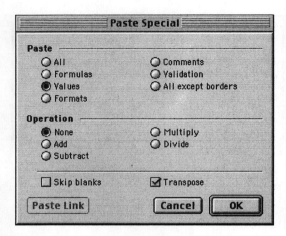

Figure I.10: Paste Special

most commonly used in this book are **transpose** check box, which transposes rows and columns, and the **Paste Values** radio button, which is useful if you need to copy a range of values defined by formulas. A straight copy will alter the cell references in the formulas and could produce nonsense. **Paste Special** solves this problem by pasting the values, not the formulas.

Filling

Suppose you need to fill cells A1:A30 with the value 1. Enter the value 1 in cell A1. Activate A1 and move the cursor to the lower right-hand corner (the **fill handle**) of A1. The cursor becomes a cross hair. Drag the fill handle and pull down to cell A30. This copies the value 1 into cells A2:A30. Alternatively, you can select A1:A30 after entering the value 1 in cell A1. Then choose **Edit − Fill − Down** from the Menu Bar. Other options are available such as **Edit − Fill − Series** if this approach is taken.

I.5   Opening Files

Often you will need to open text or data files, for instance the data files on the Student CD-ROM accompanying *IPS*. Other times, the data may be in a binary format produced by some other application.

Excel can read and open a wide selection of binary files. To see which ones may be imported, choose **File − Open** from the Menu Bar and make the appropriate selection from the drop-down list.

For text (ASCII) files, Excel may start the **Text Import Wizard**, after you make your selection from the drop-down list, once you choose **File − Open** from

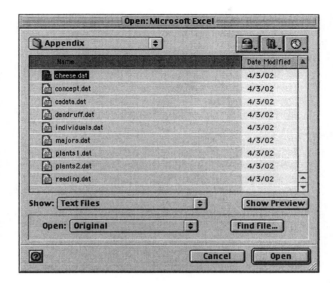

Figure I.11: Importing Files

the Menu Bar. This is a sequence of three dialog boxes specifying how the text should be imported. The **Text Import Wizard** helps you make intelligent choices and if the data file contains lines of explanatory text at the top of the file, then it is very handy for formatting correctly upon import. We illustrate using the "cheese" data set on the Student CD-ROM. Fig I.11 shows that the file "cheese.dat" has been selected following **File − Open − All Files**.

Step 1. The Text Import Wizard (Fig I.12) makes a determination that the data were **Fixed width**. You may override this with the radio button. Sometimes the Text Import Wizard interprets incorrectly. The default starting position is shown as Row 1. This too can be changed if labels are needed.

Step 2. The next screen depends on whether you chose Delimited or Fixed width in Step 1. If Delimited, you pick the delimiter. If Fixed width, you create line breaks.

Step 3. The final step allows you to select how the imported data will be formatted. Usually the radio button **General**, the default, is appropriate.

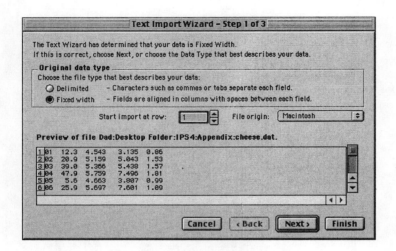

Figure I.12: Text Import Wizard—Step 1

I.6 Printing

This is generally the last step. In the **Page Setup** dialog box (Fig I.13) accessed from the Menu Bar using **File – Page Setup**, there are four tabs:

- **Page.** Orientation, scaling, paper size, print quality.
- **Margins.** Top, bottom, left, right, preview window.
- **Header/Footer.** Information printed across top or bottom.
- **Sheet.** Print area, column/row title, print order.

The **Print Preview** button on the **Standard Toolbar** or **File – Print Preview** from the Menu Bar allows you to see what portion of your document is being printed and where it is positioned. Other options are available (Fig I.14) at the top of the print preview screen to assist you in previewing your output prior to printing.

Finally, when you are satisfied with your output, press the **Print...** button at the top of the preview screen. You can also print directly with the document window open using the **Print** button on the **Standard Toolbar** or **File – Print** from the Menu Bar. This brings up a **Print** dialog box in which you select which pages to print, the number of copies, printer setup, as well as buttons for **Page Setup** and **Print Preview** just discussed.

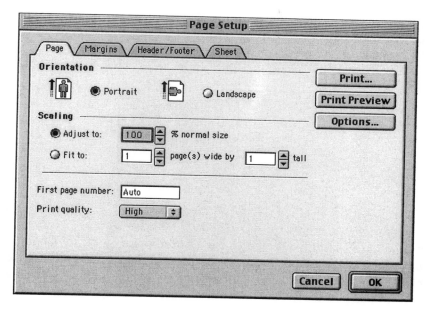

Figure I.13: Page Setup

Figure I.14: Print Preview Options

I.7 Whither?

This brief introductory chapter contains a bare-bones description of Excel, as much as you need to know to access the remainder of this book. In the following chapters, not only will you make use of many of the topics and tips presented here, but you will learn about the **ChartWizard**, enter formulas, copy and paste cells, and so on. By using Excel for statistics you will also obtain a good practical background in spreadsheets.

One way to learn more about Excel is by using the animated **Office Assistant** as in Fig I.15. You can install a gallery of different characters from the Excel installation CD-ROM in the Office:Actors folder. The one shown here is called **Max**. Max can be called up using the **Help** button (the ? at the right of the **Standard Toolbar**). Ask Max a question such as "What's new?" and see how he responds. If you **right-click (Windows)** or **Control-click (Macintosh)** Max, then more options become available.

Numerous books are available in libraries and in bookstores, but books often contain too much information. Finally, there is a wealth of recent material available

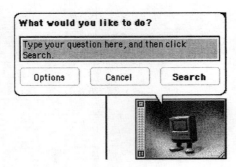

Figure I.15: Office Assistant—Max

on the Internet. Use your favorite search engine or directory and you'll find pointers to macros (Visual Basic programs) and sample workbooks made available by other users for applications of Excel. Current versions of Excel are "Internet ready" and can read html files and carry out "web queries" to import data directly from the Internet or save Excel files to html format. In fact, with some browsers and servers it is also possible for users to manipulate spreadsheets within the browser.

If you find interesting Internet resources, let me know (*hoppe@mcmaster.ca*) and I will place a pointer to them on the Excel statistics Web site, which will be located at http://www.mathematics.net/.

Chapter 1

Looking at Data–Distributions

Excel provides more than 70 functions related to statistics and data analysis as well as tools in the **Analysis ToolPak**. Additionally, the **ChartWizard** gives a step-by-step approach to creating informative graphs.

We first discuss the **ChartWizard**. The figures are from Excel 2001 for Macintosh. They differ only cosmetically from Excel 2000 for Windows, as illustrated in Fig 1.1.

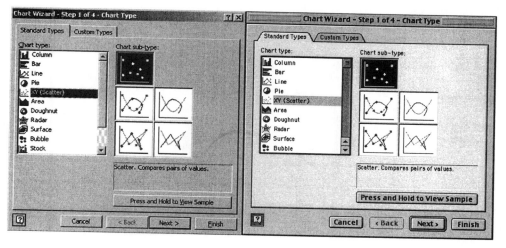

Figure 1.1: Windows 2000 and Macintosh 2001 ChartWizard Opening Screens

1.1   Displaying Distributions with Graphs

The ChartWizard

The **ChartWizard** is a step-by-step approach to creating informative graphs. Its interface provides a sequence of four steps in **Excel 97/2000 (Windows)** and **Excel 98/2001 (Macintosh)** that guide the user through the creation of a customized graph (called a Chart by Excel). The user supplies details about the chart type, formatting, titles, legends, and so on, in dialog boxes. The ChartWizard can be activated either from the button on the **Standard Toolbar** or by choosing **Insert – Chart** from the Menu Bar. The chart can be inserted in the current sheet or in a new sheet. The following applies to Excel 97/98/2000/2001. **Users of Excel 5/95 will have five steps paralleling the ones presented here.**

Example 1.1.   (Page 6 in the text.) Fig 1.2 shows the education level of young adults aged 24 to 35 given both as a count and as a percent. Create a bar chart of the percents.

	A	B	C
1	**Education**	**Count (millions)**	**Percent**
2	Less than high school	4.7	12.3
3	High school graduate	11.8	30.7
4	Some college	10.9	28.3
5	Bachelor's degree	8.5	22.1
6	Advanced degree	2.5	6.6

Figure 1.2: Educational Level

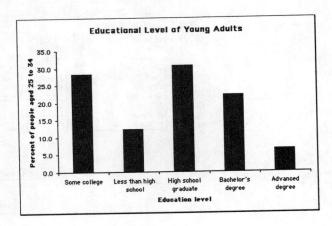

Figure 1.3: Excel Bar Chart

Solution.   Fig 1.3 is a bar graph produced by Excel that displays the same information as in Figure 1.2. For other types of graphical displays, make appropriate choices from the same sequence of dialog boxes. The following steps describe how it is obtained. First enter the data and labels in cells A1:C6 and format the display as in Figure 1.2.

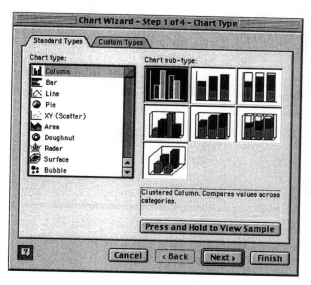

Figure 1.4: ChartWizard—Step 1

Step 1. Select cells A1:C6 and click on the **ChartWizard**. The ChartWizard (Fig 1.4) displays the types of graphs that are available. In the left field select **Column** for Chart type.

In the right field select **Clustered Column** for Chart sub-type, which is the first choice in the top row. When you select a sub-type, an explanation of the chart appears in the box below all the choices, and you can preview your chart's appearance using the "Press and Hold to View Sample" button. Click Next.

Step 2. The next dialog box (Fig 1.5) with title **Chart Source Data** previews your chart and allows you to select the data range for your chart. Since you had already selected cells A1:C6 prior to invoking the ChartWizard, this block appears in the text area **Data range**. Had you not selected the data range, then you would input the range now or you can make corrections to the data range. The preview chart shows bar charts for both the counts and the percents. To remove the counts click on the **Series** tab at the top of the dialog box, then highlight the series "Count (millions)" and click the **Remove** button (right side of Figure 1.5. The bar graph of counts vanishes. Click Next.

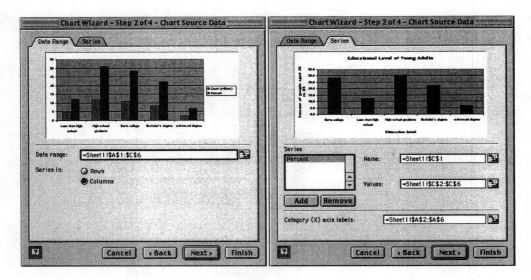

Figure 1.5: ChartWizard—Step 2

Step 3. A dialog box **Chart Options** (Fig 1.6) appears with the default chart. Rarely is the default satisfactory; you will generally need to make cosmetic changes to its appearance.

- Click the **Titles** tab. Enter "Educational Level" for Category (X) axis and "Percent of people aged 25 to 34" for Value (Y) axis.

- Click the **Legend** tab. We don't require a legend since only one variable is plotted, so make sure the check box **Show Legend** is cleared.

- Additional tabs are available to customize other types of charts. They are not required here. Click Next.

Step 4. The final step lets you decide if you want the chart placed on the same worksheet as the data or in another worksheet. With each choice there is a field for entering the worksheet name. We will embed the chart on the same worksheet, so we select the radio button **As object in:** (Fig 1.7). As our current workbook only has one sheet, Excel has used the default name Sheet1. We could also embed the chart on another sheet in the same workbook. Click the Finish button.

The chart appears with eight handles indicating that it is selected. The chart can be resized by selecting a handle and then dragging the handle to the desired size. The chart can also be moved. Click the interior of the chart and drag it to another location (holding the mouse button down). Then click outside the chart to deselect.

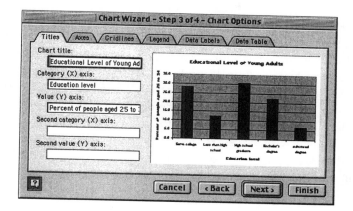

Figure 1.6: ChartWizard—Step 3

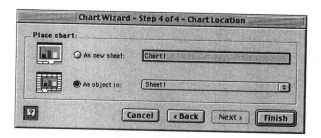

Figure 1.7: ChartWizard—Step 4

You will also find the **Chart Toolbar** (Fig 1.8) embedded on the worksheet. This is used for embellishments of the chart. Use of this toolbar is described in the next section on creating histograms. Note that the Chart Toolbar may also be called from the Menu Bar by **View – Toolbars – Chart**. Also, if you select the Chart by clicking once within its area, the Menu Bar will change. In place of the word **Data** there will now appear **Chart** from which a pull-down menu will provide the same tools as are displayed with icons on the Chart Toolbar.

Note: You might get an error message "Cannot add chart to shared workbook" even if you are not sharing your workbook. This is a bug introduced when Shared Workbooks were implemented in Excel 97/98. It occurs under certain conditons

Figure 1.8: Chart Toolbar

if you try to create a chart using the data analysis tools.

You can still output your chart to a new workbook, and then, if desired, copy the chart to the existing workbook. This is one workaround. Fortunately, this problem can be fixed by installing an updated file ProcDBRes to replace the existing one of the same name in the folder/directory "Microsoft Office 98:Office:Excel Add-Ins:Analysis Tools" (the folder for a default installation—your location may differ). This file is available for download at the Microsoft Software Library. There was a similar problem with Excel 97 (fixed in Excel 2000). Details may be found at the URLs http://support.microsoft.com/default.aspx?scid=kb;EN-US;q183188 and http://support.microsoft.com/default.aspx?scid=kb;en-us;Q178243.

Pie Charts

It is easy as pie to produce a chart as in Fig 1.9 using Excel. Select Pie in place of Column in Step 1 of the ChartWizard and follow the remaining steps.

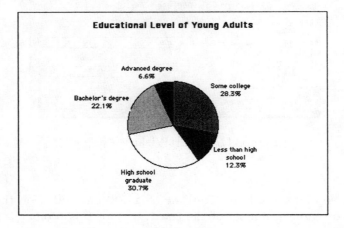

Figure 1.9: Pie Chart

Alternatively, since we have already created a bar graph, it is instructive to use the **Chart Toolbar** as an illustration of how easily modifications may be made. This interface is a vast improvement over the previous version of Excel.

1. Select the completed bar graph by clicking once within its border.

2. From the Menu Bar select **Chart – Chart Type....** You will be presented with a box that is identical to Fig 1.4 but for the title, which contains only the words **Chart Type** without mention of Step 1 of the ChartWizard. In the left field, referring to Figure 1.4, select **Pie** for Chart type and in the right field select **Pie** for Chart sub-type, which is the first choice in the top row.

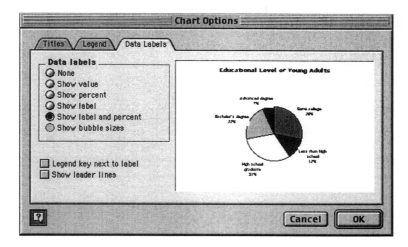

Figure 1.10: Chart Options

3. From the Menu Bar select **Chart − Chart Options....** Fig 1.10 appears with three tabs: Titles, Legend, and Data Labels. Click the tab **Data Labels** and then select the radio button **Show label and percent** and check the box **Show leader lines**. Click OK. A pie chart now replaces the bar graph.

4. Double-clicking on any label will bring up the **Format Data Label** dialog box which allows you to change background colors, number of decimals, and other formatting features.

Histograms

The ChartWizard is designed for use with data that are already grouped, for instance, categorical variables, and can therefore be used to construct a histogram of quantitative variables that have been grouped into categories or intervals. However, for raw numerical data, Excel provides additional commands within the **Analysis ToolPak** for constructing histograms.

To determine whether this toolpak is installed, choose **Tools − Add-Ins** from the Menu Bar. The **Add-Ins** dialog box appears. Depending on whether other Add-Ins have been loaded, your box might appear slightly different. If the Analysis ToolPak box is not checked, then select it and click OK. The **Analysis ToolPak** will now be an option in the pull-down menu when you choose **Tools − Data Analysis**. Note that you can also use the Select button to add customized add-ins to complement Excel.

Histogram from Raw Data

Example 1.2. (Example 1.9, page 14 in the text.) Make a histogram of the percent of adult residents in each of the 50 states who identified themselves in the 2000 census as "Spanish/Hispanic/Latino. (Fig 1.11).

	A	B	C	D	E
1	Percent of Hispanics in the adult population, by state (2000)				
2	1.5	5.7	5.6	38.7	1.2
3	3.6	6.4	2.7	13.8	2.0
4	21.3	10.7	2.4	4.3	28.6
5	2.8	3.1	1.3	1.0	8.1
6	28.1	2.3	1.8	1.6	0.8
7	14.9	5.8	1.6	4.3	4.2
8	8.0	1.3	4.5	6.5	6.0
9	4.0	2.4	16.7	2.6	0.6
10	16.1	0.6	1.4	7.0	2.9
11	5.0	4.0	12.3	2.2	5.5

Figure 1.11: Percent of Hispanics in the Adult Population

Solution.   The data appears in Table 1.2 of the text.

1. Excel requires a contiguous block of data for the histogram tool and it is convenient to enter these 50 data in a block of ten rows by 5 columns. So enter the data in a block (cells A2:E11) with the label "Percent of hispanics in the adult population, by state (2000)".

2. From the Menu Bar choose **Tools – Data Analysis** and scroll to the choice Histogram (Fig 1.12). Click OK.

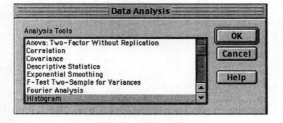

Figure 1.12: Data Analysis Tools

3. In the dialog box (Fig 1.13) type the reference for the range A2:E11 in the **Input range** area, which is the location on the workbook for the data. As with the Bar Chart, you may instead click and drag from cell A2 to E11. The choice depends on whether your preference is for strokes (keyboard) or clicks (mouse). Leave the **Bin range** blank to allow Excel to select the bins, check the **Labels** box blank, type a cell location, say A14, for **Output range** to denote the upper left cell of the output range, and check

the box **Chart output** to obtain a histogram on the same sheet of the workbook as the data. The option Pareto (sorted histogram) constructs a histogram with the vertical bars sorted from left to right in decreasing height. If Cumulative Percentage is checked, the output will include a column of cumulative percentages.

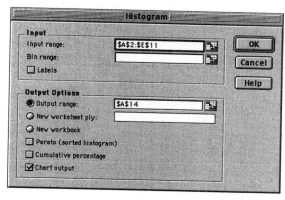

Figure 1.13: Histogram Tool

4. The output appears in Fig 1.14. The entries under "Bin" in A14:A18 are not the midpoints of the bin intervals, as you might expect. Rather they are the **upper limits** of the boundaries for each interval. The corresponding frequencies appear in cells B14:B18 with the histogram to the right. We shall shortly modify the histogram by changing the labels and allowing adjacent bars to touch. But first, we explain how to customize the selection of bins.

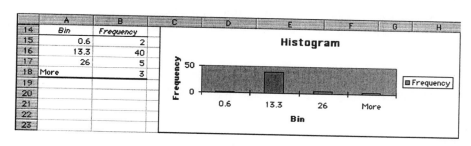

Figure 1.14: Output Table and Default Histogram

Changing the Bin Intervals

If the bin intervals are not specified, then Excel creates them automatically, choosing the number of bins roughly equal to the square root of the number of observations beginning and ending at the minimum and maximum, respectively, of the data set. In creating a histogram from raw data, we let Excel choose the default bins. Here we select our own bin intervals.

1. Type "Bin" (or another appropriate label) in an empty location, say G1. Then enter the values 5.0, 10.0, 15.0, 20.0, 25.0, 30.0, 35.0, 40.0 directly below in cells G2:G9. An easy way to accomplish this is to type 5.0 and 10.0 in cells G2 and G3 respectively, then select G2 and G3, click the fill handle in the lower right corner of G3, then drag the fill handle down to cell G9 and release the mouse button.

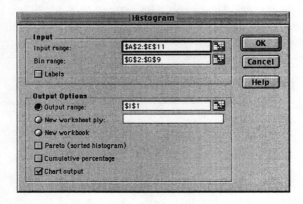

Figure 1.15: Histogram Tool—Specified Bin Intervals

2. Repeat the earlier procedure for creating a histogram, but this time type G2:G9 in the text area for **Bin range** and select a location (cell I1) in Fig 1.15) to mark where the output with your selected bin intervals will appear (see Fig 1.16).

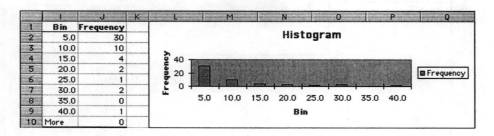

Figure 1.16: Output Table and New Bin Histogram

Enhancing the Histogram

While the default histogram captures the overall features of the data set, it is inadequate for presentation. Excel provides a set of tools for enhancing the histogram. These are too numerous for all to be mentioned here, but a few will be discussed with reference to the example. The other options may be invoked analogously.

Legend. To remove the legend (which is not needed here) select **Chart − Chart Options...** from the Menu Bar, click the **Legend** tab, and clear the box **Show legend**.

Resize. Both the histogram (called the **Plot Area**) and the box (called the **Chart Area**) that contains it can be resized and moved. Select the Chart Area by clicking once within its boundary, resize using any of the eight handles that appear, or move it by dragging or cutting and pasting from **Edit** on the Menu Bar to a new location. Likewise, select the Plot Area by clicking once within its boundary and then resize or move as with the Chart Area. The X axis labels may appear placed by default horizontally, vertically or diagonally to accommodate the selected size. This can also be changed. After removing the Legend and resizing click **outside** the Chart Area to deselect.

Bar Width. Adjacent bars do not touch in the default histogram, which looks more like a bar chart for categorical data. To adjust the bar width, click and select any one of the bars, and then from the Menu Bar select **Format − Selected Data Series...** to bring up the **Format Data Series** dialog box. Select the Options tab (Fig 1.17) and change the Gap width from 150% to 0%.

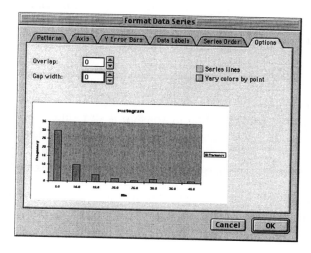

Figure 1.17: Format Data Series

Chart Title. Click on the title word Histogram. A rectangular grey border with handles will surround the word, indicating that it is selected for editing. Begin typing "Percent of Hispanics in an Adult Population," hold down the **Alt** key (**Windows**) or the **Command** key (**Macintosh**), and press Enter. You may now type a second line of text in the **Formula Bar** entry area. Continue typing "by state (2000)," and then press Enter. If you want to move the title within the Chart Area, use the handles. To change the font of your title, select the title, then from the Menu Bar choose **Format – Selected Chart Title....** The dialog box has three tabs: Patterns, Font, and Alignment. Select the **Font** tab and pick a font face, style, and size.

X axis Title. Click on the word Bin at the bottom of the chart and type "Percent of Hispanics." Change the font by selecting the X axis title, then from the Menu Bar choose **Format – Selected Axis Title...** and complete the dialog box as desired in the same fashion as for the Chart Title.

Y axis Title. Click on the word Frequency on the left side, and then from the Menu Bar choose **Format – Selected Axis Title...** for any desired formatting.

X axis Format. Double-click the X axis, and in the **Format Axis** dialog box you can click on various tabs to change the appearance of the X axis. If you click on the **Alignment** tab you can change the orientation of the X axis labels.

Y axis Format. Double-click the Y axis, and in the **Format Axis** dialog box you can click on various tabs to change the appearance of the Y axis. Click on the **Scale** tab and change the **Maximum** to 30. Sometimes if you resize the chart you will need to experiment with the **Major** or **Minor** units to achieve a pleasing result. Click OK.

More Interval. The "More" interval with 0 counts is unattractive, especially if it appears on your graph (which it may, depending on your default settings, prior to enhancement). In the workbook in cell I10 (refer to Fig 1.16), change the label "More" to 45.0. The histogram is dynamically linked to the data in columns I and J and the label "More" on the X axis becomes 45.0. Of course, knowing that the count is zero in the bin with value 45.0, we could redo the histogram ignoring this bin, if we wished to exclude it.

Plot Area Pattern. The default histogram has a border around the Plot Area and the Plot Area is shaded grey. Both defaults can be changed by double-clicking the **Plot Area** to bring up the **Format Plot Area** dialog box. If desired, select the radio button **None** for Border and also select the radio button **None** for Area.

At the conclusion of the formatting, the histogram will look like Fig 1.18.

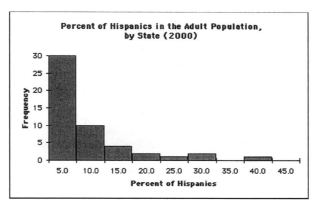

Figure 1.18: Final Histogram after Editing

Histogram from Grouped Data

The **Histogram** tool requires the raw data as input. When numerical data have already been grouped into a frequency table, it is the **ChartWizard** that is the appropriate tool. First use it to obtain a bar chart, and then modify it exactly as you would enhance a histogram.

> **Example 1.3.** (See Exercise 1.17, page 25 in the text.) Fig 1.19 gives the frequencies of vocabulary scores of all 947 seventh graders in Gary, Indiana, on the vocabulary part of the Iowa Test of Basic Skills. Column A is the interval for the score and column B is the score. Construct a histogram using the **ChartWizard**. The final histogram should be similar to that shown in Fig 1.20.

	A	B	C
1	Score	Bin	Number of Students
2	2.0 - 2.9	3	9
3	3.0 - 3.9	4	28
4	4.0 - 4.9	5	59
5	5.0 - 5.9	6	165
6	6.0 - 6.9	7	244
7	7.0 - 7.9	8	206
8	8.0 - 8.9	9	146
9	9.0 - 9.9	10	60
10	10.0 - 10.9	11	24
11	11.0 - 11.9	12	5
12	10.0 - 12.9	13	1
13	Total		947

Figure 1.19: Vocabulary Scores—Grouped Data

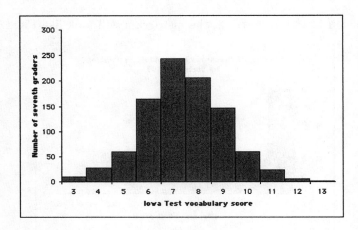

Figure 1.20: Histogram from Grouped Data

1.2   Describing Distributions with Numbers

The most direct way to obtain the common summary statistics is through the **Descriptive Statistics Tool**, which provides preformatted output very quickly. It is explained in this section. An alternative is the **Formula Palette** which provides greater flexibility of output and many more functions and formulas over its predecessor. We first describe the Descriptive Statistics Tool and then the Formula Palette.

The Descriptive Statistics Tool

> **Example 1.4.** (Based on problems 1.55 - 1.57, pages 59 - 60 in the text.)   Fig 1.21 shows the calories and sodium levels measured in three types of hot dogs: beef, meat (mainly pork and beef), and poultry. Data is from *Consumer Reports*, June 1986, pp. 366-367. Describe the data using the Descriptive Statistics Tool.

Solution.   For illustration purposes we consider only the beef calories data.

1. From the Menu Bar choose **Tools − Data Analysis** and double-click **Descriptive Statistics** (or, equivalently, select **Descriptive Statistics** and click OK) in the **Data Analysis Dialog** box. A dialog box **Descriptive Statistics** appears which prompts for user input.

2. Complete the input as follows. The **Input range:** is A3:A23, corresponding to the beef calories, including labels. (If you selected this range prior to invoking Descriptive Statistics, it will already be inserted by Excel.) Check the box **Labels in first row**. The **Confidence level for mean:** is not

	Beef hot dogs		Meat hotdogs		Poultry hotdogs	
	A	B	C	D	E	F
1	Calories	Sodium	Calories	Sodium	Calories	Sodium
4	186	495	173	458	129	430
5	181	477	191	506	132	375
6	176	425	182	473	102	396
7	149	322	190	545	106	383
8	184	482	172	496	94	387
9	190	587	147	360	102	542
10	158	370	146	387	87	359
11	139	322	139	386	99	357
12	175	479	175	507	170	528
13	148	375	136	393	113	513
14	152	330	179	405	135	426
15	111	300	153	372	142	513
16	141	386	107	144	86	358
17	153	401	195	511	143	581
18	190	645	135	405	152	588
19	157	440	140	428	146	522
20	131	317	138	339	144	545
21	149	319				
22	135	298				
23	132	253				

Figure 1.21: Hot Dog Data

needed at this time (it gives the half-width). Check the **Kth largest:** or **Kth smallest:** boxes if needed. We have selected K = 5 for illustration.b

3. The **Output Options** tell Excel where to place the output. Select cell A26. Finally check the box Summary Statistics and click OK. The output appears in Fig 1.22. We have formatted the output by reducing the number of decimal points using the **Decimal** button in the **Formatting Toolbar**. We can read off the summary statistics:

mean = 156.85
standard deviation = 5.0629
median = 152.50
minimum = 111   maximum = 190
5th smallest = 139   5th largest = 181

Formula Palette

The **Function Wizard** was replaced by the **Formula Palette** in **Excel 97/98**. This is a tool that assists in entering formulas and functions included in Excel, particularly complex ones. The functions can perform decision-making, action-taking, or value-returning operations. The Formula Palette simplifies this process by guiding you step by step.

It can be fired up in one of two ways. When you select a cell and press the

	A	B
26	*Calories*	
27		
28	Mean	156.85
29	Standard Error	5.0629
30	Median	152.50
31	Mode	149
32	Standard Deviation	22.6420
33	Sample Variance	512.6605
34	Kurtosis	-0.813
35	Skewness	-0.031
36	Range	79
37	Minimum	111
38	Maximum	190
39	Sum	3137
40	Count	20
41	Largest(5)	181
42	Smallest(5)	139

Figure 1.22: Descriptive Statistics Output

Paste Function button f_x next to the autosum button Σ on the **Standard Toolbar** (or, equivalently, choose **Insert – Function...** from the Menu Bar), an equal sign (=) appears both in the cell and in the **Formula Bar.** The **Paste Function** dialog box (Fig 1.23) appears showing all available functions grouped by category on the left and the function name on the right. Both lists have scroll bars for choices not directly visible on the screen. At the bottom of the box, the selected function is shown with the arguments it takes and a brief description. (In previous versions of Excel, a similar dialog box called **Function Wizard – Step 1 of 2** appeared.) When you click OK in the **Paste Function** box, the **Formula Palette** box appears below the **Formula Bar**, requesting parameters and an input range for the function you selected. In addition, the Formula Bar is now activated showing the Formula Palette's drop-down list control with the 10 most recently used functions, and an equal (=) sign appears in the Formula Bar showing the selected function partially constructed and awaiting completion of its arguments. You may enter these either directly into the Formula Bar or in the Formula Palette box.

The **Formula Palette** is usually invoked in a second, more direct way. Select a cell and press (=) on the **Formula Bar** to open the **Formula Palette** dialog box (Fig 1.24). On the far left side of the **Formula Toolbar** is a button with the most recently used function, in this case SUM. If this is the function you need then click on the word SUM and the **Formula Palette** dialog box will expand requesting the required parameter or the data range for the function (which can be typed directly or *referenced* by using the mouse to point to the data by clicking and dragging over cells in the data range). As you input this information, Excel will correspondingly build the function both in the **Formula Bar** and in the cell you had selected in the workbook. When you have completed entering the requested input, click OK to complete the function. If you want some other function than the default, click the small arrow in the box to the right of the function name

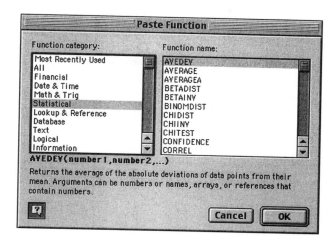

Figure 1.23: Paste Function

Figure 1.24: Formula Palette—Default

to generate a drop-down list of your 10 most recently used functions or you may select **More functions....** If you select the latter, then the **Paste Function** dialog box appears. An OK (checkmark symbol) and a cancel (X) button appear to the right of this arrow. Click the checkmark and the formula is entered into the active cell. Click the cancel to discard the formula without making changes.

Recommendation. The **Paste Function** button on the Standard Toolbar duplicates the actions of the **Formula Palette**. Since Excel formulas start with an equal (=) sign, we recommend that you begin your formulas by pressing the (=) symbol on the Formula Toolbar instead of using the Paste Function. This activates the **Formula Palette**, and you can either type the formula by hand into the Formula Bar or order up a function from the **Paste Function** box, if required. Experienced users of Excel often **customize** the **Standard Toolbar** and replace the Paste Function button with some other one.

The Five-Number Summary

Example 1.4 continued. Find the five-number summary {minimum, first quartile, median, third quartile, maximum} for the calorie distribution of the beef hot dogs data shown in Table 1.9 page 59 of the text.

	A	B	C	D
1	**Beef hot dogs**		**Five-number summary**	
2				
3	**Calories**	Min	=MIN(A4:A23)	111
4	186	Q1	=QUARTILE(A4:A23,1)	140.5
5	181	Med	=MEDIAN(A4:A23)	152.5
6	176	Q3	=QUARTILE(A4:A23,1)	140.5
7	149	Max	=MAX(A4:A23)	190
8	184			
9	190			
10	158			
11	139			
12	175			
13	148			
14	152			
15	111			
16	141			
17	153			
18	190			
19	157			
20	131			
21	149			
22	135			
23	132			

Figure 1.25: Five-Number Summary

Solution

1. Copy the beef calories data from Fig 1.21 onto a new worksheet and enter the labels "Min," "Q1," "Med," "Q3," and "Max" as shown in cells B3:B7 of your worksheet (Fig 1.25).

2. Click the equal (=) symbol on the **Formula Bar** to start the **Formula Palette** and use the drop-down list to select **More functions....** In the **Paste Function** dialog box, select **Statistical** from the left and scroll down and select QUARTILE on the right. Click OK.

3. The **Formula Palette** dialog box appears. Move it out of the way and enter the data **Array** by selecting cells A4:A23 with your mouse (or more mundanely by typing A4:A23 into the dialog box). Click in the text area for **Quart** and type "1" to indicate the first quartile. The completed formula appears in the **Formula Toolbar** and the value of the formula 140.5 shows in the dialog box (Fig 1.26). Click OK and the value 140.5 is printed in C5.

4. Continue in this fashion using the Formula Palette to complete the five number summary. Of course you can still enter the formulas by hand in the Formula Bar once you are familiar with them. Cells C4:C7 show the syntax while the values are in D4:D7.

The five-number summary is {111, 140.5, 152.5, 177.25, 190}. Note that Excel uses a slightly different definition of quartiles for a finite data set than the text.

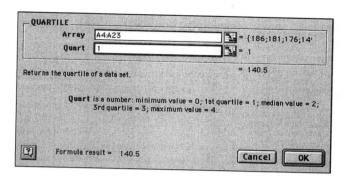

Figure 1.26: Quartile Formula

1.3   The Normal Distribution

Areas under a normal curve can be found using the NORMDIST function. The syntax is $=$ NORMDIST$(x, \mu, \sigma, \text{cumulative})$, where μ is the mean and σ is the standard deviation. The parameter cumulative indicates whether the density (set cumulative $=$ "false" or "0") or whether the cumulative distribution (set cumulative $=$ "true" or "1") is wanted. The formula $=$ NORMDIST$(x, \mu, \sigma, 1)$ returns $F(x)$, which is the area to the left of x under an $N(\mu, \sigma)$ density and can be used to produce a table of normal areas as found in many statistics texts. Another formula, $=$ NORMINV(p, μ, σ), returns the inverse $F^{-1}(p)$ of the cumulative, that is a value x such that the area to the left of x is the specified p. For $N(0,1)$, use NORMSDIST and NORMSINV instead.

Normal Distribution Calculations

Example 1.5.   (Example 1.26, page 75 in the text.) The NCAA requires Division I athletes to score at least 820 on the combined mathematics and verbal parts of the SAT exam in order to compete in their first year college year. In 2000, the scores of the more than one million students taking the SATs were approximately normal with mean 1019 and standard deviation 209. What percent of all students had SAT scores of at least 820?

Solution

Click on a cell (activate it) where you want to locate the answer, say A1. Since we are looking for the area to the right of the point 820, the syntax for $\Phi(x)$ is

$$= 1 - \text{NORMDIST}(x, \mu, \sigma, 1)$$

so enter the formula $= 1 - \text{NORMDIST}(820, 1019, 209, 1)$ since the upper tail area is wanted. The answer 0.829 appears in cell A1. (Users of **Excel 5/95** may also use the **Function Wizard** while users of **Excel 97/98** may use the **Formula Palette** instead to enter the formula.)

>   **Example 1.6.**   (Example 1.27, page 76 in the text.) The NCAA considers a student a "partial qualifer" eligible to practice and receive an athletic scholarship, but not to compete, if the combined SAT score is at least 720. What percent of all students who take the SAT would be partial qualifiers?

Solution

The syntax is $=$ NORMDIST$(820, 1019, 209, 1)$ $-$ NORMDIST$(720, 1019, 209, 1)$ which gives 0.094 for the area between 720 and 820.

>   **Example 1.7.**   (Example 1.28, page 77 in the text.) Scores on the SAT verbal test in recent years follow approximately the $N(505, 110)$ distribution. How high must a student score in order to place in the top 10% of all students taking the SAT?

Solution

We are looking for the SAT score with area 0.1 to its right under the normal curve, equivalently with area 0.9 to its left. The Excel formula for the inverse of the cumulative normal distribution is

$$\Phi^{-1}(x) = \text{NORMINV}(p, \mu, \sigma)$$

so enter $=$ NORMINV$(0.90, 505, 110)$ and read off 645.97, the 90th percentile of the SAT scores.

Graphing the Normal Curve

By combining the **ChartWizard** and the NORMDIST function, we can create a graph of any normal curve. *In fact, the procedure described here can be used to plot the graph of any function Excel can evaluate.*

Constructing a Graph of the Standard Normal Curve

1. Enter the labels z and $f(z)$ in cells A2, B2. (See Fig 1.27.) $f(z)$ will represent the density or height of the normal curve at the point z.

2. Enter -3.5 and -3.4 in cells A3 and A4, respectively. These two points mark the ends of the range of values over which the normal density will be graphed. Next we create a column of z values at which the standard normal density will be calculated. Select A3:A4, check the fill handle in the lower right corner of A4, and drag to cell A73 to fill the column with decreasing values of z decremented by 0.1. Format the values with two decimal places.

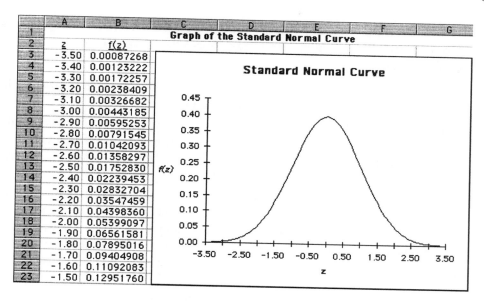

Figure 1.27: Graphing a Normal Density

3. Select cell B3 and enter $=$ NORMDIST(A3, 0, 1, 0) in the **Formula Bar**. Cell B3 now contains the value 0.00087268, the standard normal density evaluated at $z = -3.50$.

4. Select cell B3, click the fill handle, and drag down to B73. This copies the formula you just entered in B3 into cells B4:B73 relative to the corresponding cell references in column A. Column B is filled with values $f(z)$ of the standard normal density corresponding to each value of z in column A.

5. **Users of Excel 5/95.** Select cells A2:B73, click the **ChartWizard** button, and then click in cell C2 and drag to I26 to locate the graph. A dialog box **ChartWizard – Step 1 of 5** appears.

   - In Step 1 you are given an opportunity to correct or confirm your range.

   - In Step 2 select **XY (Scatter)** chart.

   - In Step 3 select format **6**.

   - In Step 4 select the radio button **Columns** for Data Series, enter "1" for Column for Category (X) Axis Labels, and enter "1" for Row for Legend Text.

   - In Step 5 select the radio button **No** for Add a Legend?, type "Standard Normal Curve" as the Chart Title, and type z and $f(z)$ for Category (X) and Value (Y) titles, respectively. Finally, click Finish.

Users of Excel 97/98/2000/2001. Select cells A2:B73 and click the **ChartWizard.** A dialog box **ChartWizard – Step 1 of 4 – Chart Type** appears.

- In Step 1 select **XY (Scatter)** for Chart Type and the lower right Chart sub-type **Scatter without markers.**

- In Step 2 under the **Data Range** tab, the range will already be indicated and the Series radio button for **Columns** will be selected. You may edit the range if it is incorrect. Under the **Series** tab no changes are necessary.

- In Step 3 under the **Titles** tab, type "Standard Normal Curve" as the Chart Title, z for Value (X) Axis, and $f(z)$ for Value (Y) Axis. Under the **Axes tab**, both check boxes should be selected. Under the **Gridlines** tab, clear all check boxes. Under the **Legend** tab, clear the Show legend. Finally, under the **Data Labels** tab, select the radio button **None.**

- In Step 4 embed the graph in the current workbook by selecting the radio button **As object in.** Finally, click Finish.

6. Activate the graph for editing and format the display as you wish to present it using the editing features discussed previously.

Constructing a Normal Table

It is very easy in Excel to produce a table of normal areas. This method described below can be adapted to produce tables of other continuous distributions.

Fig 1.28 gives areas under a standard normal curve for values of $z \geq 0$. Here is the procedure for producing this table.

1. Enter the label and values in column A and row 3. Column A is the first decimal while row 1 is the second decimal of z .

2. Enter the formula =NORMDIST($A4 + B$3) in cell B4. (Remember that the $ sign prefix makes the corresponding row or column label absolute.) Select cell B4, click the fill handle in the lower right corner of B4, and drag to K4.

3. Select cells B4:K4, click the fill handle in the lower right corner of K4, and drag to K43 to fill the block B4:K43.

Normal Quantile Plots

Excel does not provide a normal quantile (probability) plot, but it is very easy to construct such a graph. We defer this to the next chapter since it is an application of a scatterplot.

	A	B	C	D	E	F	G	H	I	J	K
1					Normal Table Constructed in Excel						
2											
3	z	0.00	0.01	0.02	0.03	0.04	0.05	0.06	0.07	0.08	0.09
4	0.00	0.5000	0.5040	0.5080	0.5120	0.5160	0.5199	0.5239	0.5279	0.5319	0.5359
5	0.10	0.5398	0.5438	0.5478	0.5517	0.5557	0.5596	0.5636	0.5675	0.5714	0.5753
6	0.20	0.5793	0.5832	0.5871	0.5910	0.5948	0.5987	0.6026	0.6064	0.6103	0.6141
7	0.30	0.6179	0.6217	0.6255	0.6293	0.6331	0.6368	0.6406	0.6443	0.6480	0.6517
8	0.40	0.6554	0.6591	0.6628	0.6664	0.6700	0.6736	0.6772	0.6808	0.6844	0.6879
9	0.50	0.6915	0.6950	0.6985	0.7019	0.7054	0.7088	0.7123	0.7157	0.7190	0.7224
10	0.60	0.7257	0.7291	0.7324	0.7357	0.7389	0.7422	0.7454	0.7486	0.7517	0.7549
11	0.70	0.7580	0.7611	0.7642	0.7673	0.7704	0.7734	0.7764	0.7794	0.7823	0.7852
12	0.80	0.7881	0.7910	0.7939	0.7967	0.7995	0.8023	0.8051	0.8078	0.8106	0.8133
13	0.90	0.8159	0.8186	0.8212	0.8238	0.8264	0.8289	0.8315	0.8340	0.8365	0.8389
14	1.00	0.8413	0.8438	0.8461	0.8485	0.8508	0.8531	0.8554	0.8577	0.8599	0.8621
15	1.10	0.8643	0.8665	0.8686	0.8708	0.8729	0.8749	0.8770	0.8790	0.8810	0.8830
16	1.20	0.8849	0.8869	0.8888	0.8907	0.8925	0.8944	0.8962	0.8980	0.8997	0.9015
17	1.30	0.9032	0.9049	0.9066	0.9082	0.9099	0.9115	0.9131	0.9147	0.9162	0.9177
18	1.40	0.9192	0.9207	0.9222	0.9236	0.9251	0.9265	0.9279	0.9292	0.9306	0.9319
19	1.50	0.9332	0.9345	0.9357	0.9370	0.9382	0.9394	0.9406	0.9418	0.9429	0.9441
20	1.60	0.9452	0.9463	0.9474	0.9484	0.9495	0.9505	0.9515	0.9525	0.9535	0.9545
21	1.70	0.9554	0.9564	0.9573	0.9582	0.9591	0.9599	0.9608	0.9616	0.9625	0.9633
22	1.80	0.9641	0.9649	0.9656	0.9664	0.9671	0.9678	0.9686	0.9693	0.9699	0.9706
23	1.90	0.9713	0.9719	0.9726	0.9732	0.9738	0.9744	0.9750	0.9756	0.9761	0.9767
24	2.00	0.9772	0.9778	0.9783	0.9788	0.9793	0.9798	0.9803	0.9808	0.9812	0.9817
25	2.10	0.9821	0.9826	0.9830	0.9834	0.9838	0.9842	0.9846	0.9850	0.9854	0.9857
26	2.20	0.9861	0.9864	0.9868	0.9871	0.9875	0.9878	0.9881	0.9884	0.9887	0.9890
27	2.30	0.9893	0.9896	0.9898	0.9901	0.9904	0.9906	0.9909	0.9911	0.9913	0.9916
28	2.40	0.9918	0.9920	0.9922	0.9925	0.9927	0.9929	0.9931	0.9932	0.9934	0.9936
29	2.50	0.9938	0.9940	0.9941	0.9943	0.9945	0.9946	0.9948	0.9949	0.9951	0.9952
30	2.60	0.9953	0.9955	0.9956	0.9957	0.9959	0.9960	0.9961	0.9962	0.9963	0.9964
31	2.70	0.9965	0.9966	0.9967	0.9968	0.9969	0.9970	0.9971	0.9972	0.9973	0.9974
32	2.80	0.9974	0.9975	0.9976	0.9977	0.9977	0.9978	0.9979	0.9979	0.9980	0.9981
33	2.90	0.9981	0.9982	0.9982	0.9983	0.9984	0.9984	0.9985	0.9985	0.9986	0.9986
34	3.00	0.9987	0.9987	0.9987	0.9988	0.9988	0.9989	0.9989	0.9989	0.9990	0.9990
35	3.10	0.9990	0.9991	0.9991	0.9991	0.9992	0.9992	0.9992	0.9992	0.9993	0.9993
36	3.20	0.9993	0.9993	0.9994	0.9994	0.9994	0.9994	0.9994	0.9995	0.9995	0.9995
37	3.30	0.9995	0.9995	0.9995	0.9996	0.9996	0.9996	0.9996	0.9996	0.9996	0.9997
38	3.40	0.9997	0.9997	0.9997	0.9997	0.9997	0.9997	0.9997	0.9997	0.9997	0.9998
39	3.50	0.9998	0.9998	0.9998	0.9998	0.9998	0.9998	0.9998	0.9998	0.9998	0.9998
40	3.60	0.9998	0.9998	0.9999	0.9999	0.9999	0.9999	0.9999	0.9999	0.9999	0.9999
41	3.70	0.9999	0.9999	0.9999	0.9999	0.9999	0.9999	0.9999	0.9999	0.9999	0.9999
42	3.80	0.9999	0.9999	0.9999	0.9999	0.9999	0.9999	0.9999	0.9999	0.9999	0.9999
43	3.90	1.0000	1.0000	1.0000	1.0000	1.0000	1.0000	1.0000	1.0000	1.0000	1.0000

Figure 1.28: Normal Table

1.4 Boxplots

A boxplot is one of the most important exploratory tools available to the data analyst. Unfortunately this tool is not part of the Analysis ToolPak. The Microsoft Personal Support Center does have a Web page "XL: How to Create a BoxPlot – Box and Whisker Chart (Q155130)" located at

$$http: //support.microsoft.com/support/kb/articles/q155/1/30.asp$$

with instructions for creating a reasonable boxplot using a **Volume-Open-High-Low-Close** Chart but the sequence of steps is complicated enough to discourage a student from using this tool regularly. We have provided a macro *boxplot.xls* which will produce a boxplot of a single data set or side-by-side boxplots of multiple data sets. This boxplot can be downloaded from the Freeman Web site.

Example 1.8.   (See Exercise 1.49, page 57 in the text.) Fig 1.29 shows the scores on the Survey of Study Habits and Attitudes for 18 first-year college women. Find the five-number summaries for both sets of SSHA scores and make side-by-side modified boxplots for the two distributions.

	A	B	C	D	E
1				Women	Men
2				154	108
3				109	140
4		Run Boxplot		137	114
5				115	91
6				152	180
7				140	115
8				154	126
9				178	92
10				101	169
11				103	146
12				126	109
13				126	132
14				137	75
15				165	88
16				165	113
17				129	151
18				200	70
19				148	115
20					187
21					104

Figure 1.29: Boxplot Input

Solution.   Prepare the data in columns on a worksheet as in Fig 1.29. The data need not be in adjacent columns nor need the number of observations in each data set be the same. If the data are on the same worksheet as the boxplot macro then click the button **Run Boxplot** to bring up the Boxplot dialog box Fig 1.30 where you select the data. If the data sets have the same column lengths

Figure 1.30: Boxplot Dialog Box

then an entire block can be selected. Otherwise each column needs to be selected separately. In this example, first select cells D1:D19, then hold down the **Control (Windows)** or **Command (Macintosh)** keys and select E1:E21. Check the box "First Row Contains Labels." You can also label the data axis (Y-axis) if desired. The boxplot, Fig 1.30, will appear in the next worksheet, named "Box Plot 1," in the same workbook together with the five number summaries for the data sets. Outliers are also plotted.

The data need not be on the same sheet as the boxplot button as long as the boxplot.xls file is open. If the data is in another workbook then the macro is run as follows. From the Menu Bar select **Tools − Macro...** to open the **Macro** dialog box and select **Run** which opens the Boxplot dialog box. Complete this box as before.

Fig 1.31 shows the resulting boxplot together with the five number summary. Outlier values are also printed indicating that the value 200 among the women's scores is an outlier according to the $1.5 \times IQR$ rule.

Note that Excel uses a slightly different definition for the quartiles than is given in the text. For the women's scores the quartiles calculated are the same but not for the men's scores. This may affect judgment on which observations would be considered outliers.

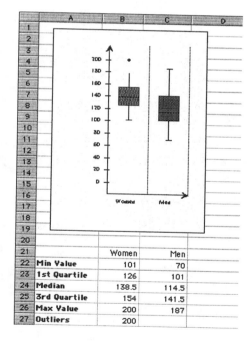

		Women	Men
22	Min Value	101	70
23	1st Quartile	126	101
24	Median	138.5	114.5
25	3rd Quartile	154	141.5
26	Max Value	200	187
27	Outliers	200	

Figure 1.31: Boxplot Output

Chapter 2

Looking at Data–Relationships

In many studies, both independent and dependent variables typically arise either as part of a controlled experiment or as an observational study. Before any specific model is imposed that can be tested statistically, it is important to judge graphically whether any relationship is justified.

If we let x denote the explanatory (independent) variable and y the response (dependent) variable, then we might plot in Cartesian coordinates all pairs (x_i, y_i) of observed values. This is called a **Scatterplot**.

2.1 Scatterplots

The steps involved in creating a scatterplot are similar to those for producing a **Histogram** using the **ChartWizard**. The instructions are based on **Excel 97/98/2000/2001. Excel 5/95** users should make corresponding changes.

Table 2.1: Fuel Consumption and Speed

Speed	10	20	30	40	50	60	70	80
Fuel	21.00	13.00	10.00	8.00	7.00	5.90	6.30	6.95

Speed	90	100	110	120	130	140	150
Fuel	7.57	8.27	9.03	9.87	10.79	11.77	12.83

Example 2.1. (Exercise 2.10, page 122 in the text.) How does the fuel consumption of a car change as its speed increases? Table 2.1 gives data for a British Ford Escort. Speed is measured in kilometers per hour, and fuel consumption is measured in liters of gasoline used per 100 kilometers travelled. Make a scatterplot with speed on the horizontal axis.

Creating a Scatterplot

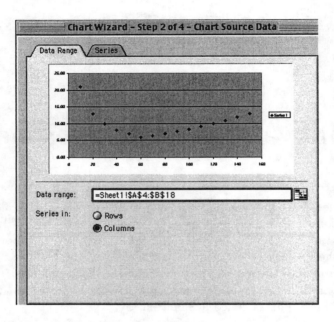

Figure 2.1: ChartWizard—Step 2

Enter the data from Table 2.1 into consecutive columns of a worksheet (for instance in cells A4:B18 and as shown in Fig 2.3 later in this section).

Step 1. Select cells A4:B22 and click on the **ChartWizard**. From the choice of charts select **XY (Scatter)** for Chart type on the left and select the top Chart sub-type **Scatter** on the right. Click Next.

Step 2. The next dialog box previews the chart and allows any changes to be made to the data range (see Fig 2.1). Click Next.

Step 3. The **Chart Option** dialog box appears (Fig 2.2).

- Click the **Titles** tab and enter "Fuel vs. Speed" for Chart title, "Speed (km/hr)" for Value (X) Axis, and "Fuel used (liters)" for Value (Y) Axis.
- Click the **Legend** tab. Clear the Show Legend check box. Click Next.
- Click the **Gridlines** tab and make sure that the Major gridlines box is checked for both axes. Click Next.

Step 4. In the last step select the radio button to enter the chart as an object on the current sheet. Click Finish. The scatterplot appears embedded in your workbook (Fig 2.3).

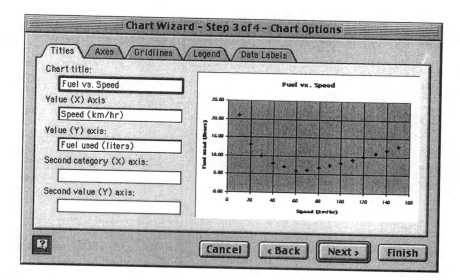

Figure 2.2: Preview of Scatterplot—Step 3

Enhancing a Scatterplot

The scatterplot may be enhanced using editing tools, some of which were described in Chapter 1. Activate the scatterplot by clicking once within its border to access new commands that become available under the Menu Bar. For instance, compare the pull-down options under **Insert**, **Format**, **Tools** as well as the new **Chart**.

Changing Scale

Excel uses a range from 0 to 100% as the default, and sometimes the scatterplot will show unwanted blank space. The scatterplot can be edited to change the maximum or minimum X axis value by double clicking the X axis to produce the **Format Axis** dialog box. Equivalently you can select the X axis by clicking once and then choose **Format** from the Menu Bar. Make any changes you wish by selecting the appropriate tabs at the top of the dialog box. If desired, the Y axis may similarly be selected for editing. Refer to the discussion for enhancing a histogram in Section 1.1, applies to any chart whether it is a histogram or a scatterplot.

Changing Titles

You can change titles on the scatterplot after you have completed it. Click on the X axis title "Speed (km/hr)" to select it and begin typing. Similarly the Y axis title and the chart title can be changed.

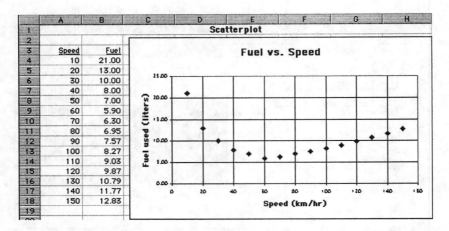

Figure 2.3: Speed and Fuel Data and Scatterplot Embedded in Sheet

Labeling a Data Point

By default Excel uses diamonds to plot the points. Suppose, for presentation purposes, you wish to use a different shape (and color) to represent a point and also to label a point. In particular suppose you wish to label the point representing minimum speed as "Min." The following steps describe how to achieve this.

1. Activate the chart and click on the minimum observation.

2. Hold down the **Control** key **(Windows)** or **Command** key **(Macintosh)** and with your mouse pointer **select** the minimum point. Release the mouse button and select the point again. The pointer becomes a four-pointed plus sign (Fig 2.4). You can now access new commands under **Format** on the Menu Bar which let you edit the selected point.

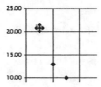

Figure 2.4: Selecting a Point

3. For **Excel 5/95** choose **Insert − Data Labels** from the Menu Bar to open the **Format Data Point** dialog box. For **Excel 97/98/2000/2001** choose **Format − Selected Data Point...** from the Menu Bar to open a corresponding **Format Data Point** dialog box. Under the **Data Labels**

tab select the radio button for Show Value. Click OK. Excel attaches the y value 21.00 to this point on the scatterplot and encloses it within a bordered selection box ready for editing. Type "min" (which appears in the **Formula Bar**) and press enter. The selection box now contains the word "min." You can select and move it and then **deselect** by clicking elsewhere.

Changing the Marker and Color of a Data Point

Sometimes you may want to make a point stand out by changing its symbol and color on the scatterplot. We show how this is done using the "min" point above.

1. Activate the chart and click on the "min" observation as before.

2. Hold down the **Control** key (**Windows**) or **Command** key (**Macintosh**) and select the point so that the pointer again becomes a four-pointed plus sign.

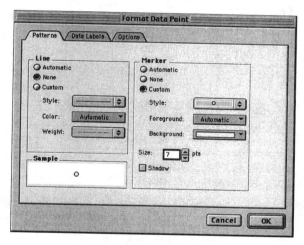

Figure 2.5: Changing the Default Marker

3. For **Excel 5/95**, choose **Insert – Data Labels** from the Menu Bar to open the **Format Data Point** dialog box. For **Excel 97/98/2000/2001**, choose **Format – Selected Data Point...** from the Menu Bar to open a corresponding **Format Data Point** dialog box. Under the **Patterns** tab, leave the **Line** selection as **None**. Under **Marker**, select a marker type from the pull-down list for **Style**, and also select a Foreground and Background color and size (Fig 2.5). In **Excel 5/95** the size of the marker cannot be changed and there is no **Options** tab. Your selection is previewed in the small **Sample** box in the lower portion of the dialog box. Click OK. Figure 2.6 shows the result of the above editing.

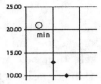

Figure 2.6: Editing a Point

Normal Quantile Plots

There are several ways to assess whether a data set is normal. An analytic approach beyond the level of this book was developed by S. Shapiro and M. B. Wilk (An analysis of variance test for normality, *Biometrika* **52**, pp. 591–611, 1965). A simple graphical approach constructs a histogram and compares the observed counts with the 68-95-99.7% rule. A more sensitive version of this idea is to order the observations and examine their distribution visually, using a scatterplot involving the corresponding expected quantiles of a normal curve. Normal data will tend to fall on a straight line. (This is the basis for the Shapiro-Wilk test.) Excel does not provide a normal quantile plot, one can easily be constructed. The expected value of the ith order statistic (the ith largest in increasing magnitude) of a sample of size n from a $N(0,1)$ distribution can be approximated by the percentile

$$z_{(i)} = \texttt{NORMSINV}\left(\frac{i - \frac{3}{8}}{n + \frac{1}{4}}\right)$$

which is the value of a standard normal such that the area to the left is $\frac{i-\frac{3}{8}}{n+\frac{1}{4}}$. So plot $z_{(i)}$ on the vertical axis against $x_{(i)}$ on the horizontal axis where $x_{(i)}$ is the ith largest from the data set $\{x_1, x_2, \ldots, x_n\}$ using the **ChartWizard**.

	A	B	C	D	E	F
1	Newcomb's Measurements					
2	of the Speed of Light					
3						
4	28	26	33	24	34	-44
5	27	16	40	-2	29	22
6	24	21	25	30	23	29
7	31	19	24	20	36	32
8	36	28	25	21	28	29
9	37	25	28	26	30	32
10	36	26	30	22	36	23
11	27	27	28	27	31	27
12	26	33	26	32	32	24
13	39	28	24	25	32	25
14	29	27	28	29	16	23

Figure 2.7: Newcomb Data

Example 2.2. (Example 1.29, page 79 in the text.) Fig 2.7 contains 66 measurements of the speed of light made by Simon Newcomb be-

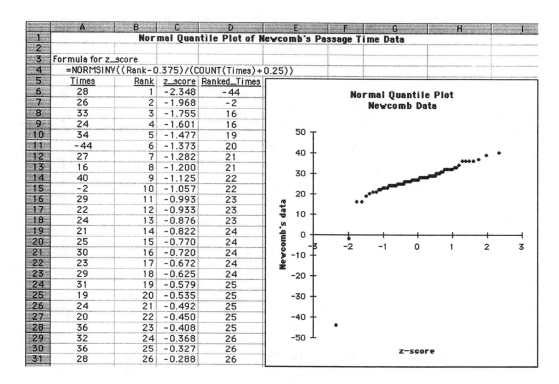

Figure 2.8: Normal Quantile Plot

tween July and September 1882. The measurements give the deviation from 24,800 nanoseconds baseline. Construct a normal quantile plot of the data.

Solution

1. Referring to columns A:D in Fig 2.8, reenter the data in cells A6:A71 of a workbook and the label "Times" in A5. From the Menu Bar, choose **Data** − **Sort** to sort the data in increasing order and enter the sorted data in D6:D71.

2. Enter the label "Rank" in B5 followed by the integers $\{1, 2, \ldots, 66\}$ in B6:B71.

3. Enter the label "z_score" in C5. Name the ranges "Times" and "Rank." Then select cell C6 and enter

$$= \texttt{NORMSINV}((\text{Rank} - 0.375)/(\texttt{COUNT}(\text{Times}) + 0.25))$$

Click the fill handle at the lower right corner of C6 and fill to C71.

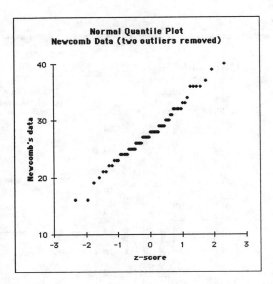

Figure 2.9: Normal Quantile Plot—Outliers Removed

4. Select a cell to locate the quantile plot, then click the **ChartWizard** button. Choose scatterplot as the chart, enter the data range C5:D71, and complete the remaining steps. Enhance the chart to reproduce the scatterplot shown in Fig 2.8 which also shows a portion of the data on the worksheet.

Two outliers are apparent $\{-44, -2\}$. Remove them from the plot and redo the calculations. The result is Fig 2.9, whose straight line appearance is an indication that a normal distribution fits the data quite well after the two outliers have been removed.

2.2   Correlation

The correlation between two variables x and y measures the strength of the linear association between them. For n pairs (x_i, y_i), $1 \leq i \leq n$, of data points the sample correlation coefficient is defined to be

$$r = \frac{1}{n-1} \sum_{i=1}^{n} \left(\frac{x_i - \bar{x}}{s_x} \right) \left(\frac{y_i - \bar{y}}{s_y} \right)$$

where $\bar{x}$ and $\bar{y}$ are the sample means of the $\{x_i\}$ and $\{y_i\}$, respectively, and s_x and s_y are the corresponding sample standard deviations.

Using the CORREL Function

The most direct way to find the correlation for Example 2.1 is by the **Formula Palette** for **Excel 97/98/2000/2001** or the **Function Wizard** for **Excel 5/95** with the function CORREL, which computes the correlation coefficient.

> **Example 2.3.** (Exercise 2.21, page 131 in the text.) Find the correlation r between the heights of the women and men in Table 2.2

Table 2.2: Heights of Women and Men

Women (x)	66	64	66	65	70	65
Men (y)	72	68	70	68	71	65

Solution

1. Enter the above data into two adjacent columns in a worksheet, say, cells A1:B6. Select an empty cell where you want the correlation to appear and invoke either the **Function Wizard** or the **Formula Palette**. In each case, select **Statistical** for Function Category and CORREL for Function Name.

2. Enter A1:A6 for **Array1** and B1:B6 for **Array2**. You may enter by hand or click and drag on the workbook over the range A1:A6, press Tab on the keyboard, then click and drag over the range B1:B6, and finally click OK (or Finish, for **Excel 5/95**). The answer 0.565 appears in the cell you selected.

Using the ToolPak

Correlation between two variables can also be calculated using the **Correlation** tool in the **Analysis ToolPak**. This tool is most effective, however, for determining pairwise correlations for multivariate data sets for which repeated use of the CORREL function would be inefficient.

This tool prints out a matrix of correlations. Such a matrix is helpful in multiple regression in deciding which variables to include in a model.

> **Example 2.4.** Darlene Gordon of the Purdue University School of Education provided the data partly shown in Fig 2.10. The data were collected on 78 seventh-grade students in a rural Midwestern school. The research concerned the relationship between the students' "self-concept" and their academic performance. The variables are OBS, a subject identification number; GPA, grade point index; IQ, score on an IQ test; AGE, age in years; SEX, female (F) or male (M); SC, overall score on the Piers-Harris self-concept scale; and C1–C6, "cluster scores"

	A	B	C	D	E	F	G	H	I	J	K	L
1			Self-concept and Academic Performance									
2	OBS	GPA	IQ	AGE	SEX	SC	C1	C2	C3	C4	C5	C6
3	1	7.9	111	13	2	67	15	17	13	13	11	9
4	2	8.3	107	12	2	43	12	12	7	7	6	6
5	3	4.6	100	13	2	52	11	10	5	8	9	7
6	4	7.5	107	12	2	66	14	15	11	11	9	9
7	5	8.9	114	12	1	58	14	15	10	12	11	6
8	6	7.6	115	12	2	51	14	11	7	8	6	9
9	7	7.7	111	13	2	71	15	17	12	14	11	10
10	8	2.4	97	13	2	51	10	12	5	11	5	6
11	9	6.0	100	13	1	49	12	9	6	9	6	7
12	10	8.8	112	12	2	51	15	16	4	9	5	8
13	11	7.5	104	12	1	35	12	5	3	2	4	7
14	12	5.5	89	13	1	54	16	8	3	11	7	7
15	13	7.2	104	13	2	54	16	14	6	7	2	7
16	14	7.6	102	13	1	64	14	12	8	10	12	9
17	15	4.7	91	14	1	56	14	13	8	10	7	8
18	16	8.2	114	13	1	69	15	15	9	12	11	9
19	17	7.8	114	13	1	55	14	11	6	11	11	9
20	18	7.6	103	12	1	65	16	16	5	12	11	9

Figure 2.10: Self-concept and Academic Performance Data Set

for specific aspects of self-concept: C1 = behavior, C2 = school status, C3 = physical, C4 = anxiety, C5 = popularity, and C6 = happiness.

Find the correlations between the response variable GPA and each of the explanatory variables IQ, AGE, SEX, SC, and C1–C6. Of all the explanatory variable, IQ does the best job of explaining GPA in a simple linear regression. How do you know this without doing all the regressions?

Solution.   Fig 2.10 shows the first 30 sets of observations in this data set, to which the following instructions apply. For the full set merely select the appropriate Input range.

1. From the Menu Bar choose **Tools − Data Analysis** and in the dialog box highlight **Correlation** and click OK.

2. In the next dialog box, **Correlation**, enter A2:L32 for **Input range** (most conveniently done by clicking and dragging over this range on the workbook and pressing the Tab key). Check the box **Labels in first row** and point to cell N1 for **Output range**. Click OK.

Excel Output

The output appears in N1:Z13, as shown in Fig 2.11. In view of symmetry, only half the correlation matrix is required. From Column P we read off the correlations between GPA and the explanatory variables. The largest correlation involving GPA is with IQ and is 0.709.

	N	O	P	Q	R	S	T	U	V	W	X	Y	Z
1		OBS	GPA	IQ	AGE	SEX	SC	C1	C2	C3	C4	C5	C6
2	OBS	1											
3	GPA	0.318	1										
4	IQ	0.520	0.709	1									
5	AGE	0.110	-0.164	-0.299	1								
6	SEX	-0.087	-0.051	0.211	-0.103	1							
7	SC	0.255	0.654	0.415	0.094	-0.098	1						
8	C1	-0.116	0.562	0.099	-0.080	-0.248	0.685	1					
9	C2	0.109	0.679	0.450	0.050	0.020	0.874	0.633	1				
10	C3	0.292	0.601	0.578	0.068	0.027	0.790	0.316	0.733	1			
11	C4	0.192	0.484	0.318	0.150	-0.021	0.857	0.474	0.787	0.662	1		
12	C5	0.230	0.590	0.399	0.071	-0.320	0.828	0.480	0.667	0.717	0.760	1	
13	C6	0.325	0.644	0.396	0.209	-0.124	0.812	0.517	0.685	0.676	0.572	0.681	1

Figure 2.11: Correlation ToolPak Output

2.3 Least-Squares Regression

We have seen how to plot two variables against each other in a scatterplot and have calculated the correlation coefficient to measure the strength of the linear association between them. It is useful to have an analytic relationship between the explanatory variable x and the response variable y of the form

$$y = f(x)$$

for predicting y from x. Such a relationship is called a simple (meaning one explanatory variable) **regression curve**. The simplest curve is a straight line

$$y = a + bx$$

called the regression line of y on x. The regression line represents, under certain assumptions, the mean response at each specified value x.

The method used to determine the coefficients a and b goes back at least to the great mathematician Gauss and is called the **Principle of Least-Squares**. Gauss himself recognized that the criterion was arbitrary and he used it because the coefficients a and b were then solvable in closed form. (Additional reasons connected with the errors being normal are presented in more advanced treatments.)

For a given x_i, we call

$$\hat{y}_i = a + bx_i$$

the **predicted** value and

$$e_i = y_i - \hat{y}_i$$

the **residual**. The **error sum of squares** is defined to be

$$\sum_{i=1}^{n} e_i^2 = \sum_{i=1}^{n} (y_i - a - bx_i)^2$$

By differentiating with respect to a and b, we can solve for the values that minimize $\sum_{i=1}^{n} e_i^2$. These are the values used in the regression line. They are given by the formulas

$$\text{slope} \quad b = r\,\frac{s_y}{s_x}$$
$$\text{intercept} \quad a = \bar{y} - b\bar{x}$$

where $\bar{x} = \frac{1}{n}\sum_{i=1}^{n} x_i$, $\bar{y} = \frac{1}{n}\sum_{i=1}^{n} y_i$, r is the correlation coefficient, and

$$(n-1)s_x^2 = \sum_{i=1}^{n}(x_i - \bar{x})^2 = \sum_{i=1}^{n} x_i^2 - \frac{1}{n}\left(\sum_{i=1}^{n} x_i\right)^2$$

$$(n-1)s_y^2 = \sum_{i=1}^{n}(y_i - \bar{y})^2 = \sum_{i=1}^{n} y_i^2 - \frac{1}{n}\left(\sum_{i=1}^{n} y_i\right)^2$$

Fitting a Line to Data

Excel provides three built-in methods for regression analysis: **Trendline**, the **Regression** tool in the **Analysis ToolPak**, and regression functions such as FORECAST and TREND. For merely graphing a regression line and providing its equation and the coefficient of determination r^2, the **Trendline** command suffices. We will consider the **Regression** tool in Chapters 10 and 11, as well as regression functions.

Linear Trendline

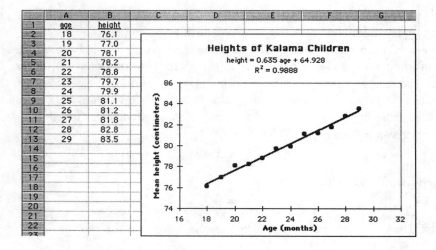

Figure 2.12: Kalama Data and Scatterplot

We use the **Linear Trendline** to insert a curve on a scatterplot. The trendline can be added to any scatterplot even after the **Regression** tool is used.

> **Example 2.5.** (Example 2.10, page 135 in the text.) Columns A, B of Fig 2.12 give the data for a group of children in Kalama, an Egyptian village that was the site of a study of nutrition in developing countries. The explanatory variable x is age, the response variable y is height. Fit the least-squares regression line to the data.

Solution. We first construct a scatterplot of the data (also shown in Fig 2.12) to verify that a linear model is appropriate.

1. Enter the data in cells A2:B13 of a workbook. Use the **ChartWizard** to create a scatterplot and edit it (primarily to change the horizontal scale), as discussed previously, so that it appears as shown in Fig 2.12. The scatterplot shows an approximate linear relationship, so it is appropriate to fit the data pairs with a straight line.

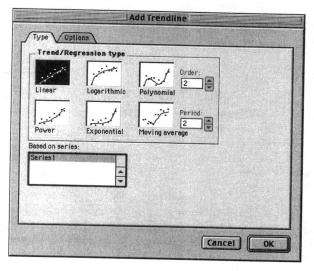

Figure 2.13: Trendline Type

2. Activate the chart for editing and select the data by **clicking on one of the points**. The points appear highlighted, the **Name** box in the **Formula Bar** shows S1, and in the text entry area we can read

$$= \mathtt{SERIES}(,\mathtt{Sheet1}!\$A\$2 : \$A\$13, \ \mathtt{Sheet1}!\$B\$2 : \$B\$13, 1)$$

meaning that the series has been selected. (Refer to the online help for more information on this function and the Introduction for the meaning of Sheet1! notation.)

3. For **Excel 5/95**, choose **Insert – Trendline** from the Menu Bar; for **Excel 97/98/2000/2001**, choose **Chart – Add Trendline...** from the Menu Bar. Then proceed as follows. Click the **Type** tab and select **Linear** (Fig 2.13). **Excel 97/98/2000/2001** have an additional text area (**Based on series**) in the blank space at the bottom of this figure. Click the **Options** tab and select the radio button **Automatic:Linear (Series1)**. Check the boxes **Display Equation on Chart** and **Display r-squared Value on Chart**. Make sure that the **Set Intercept** box is clear. Click OK. The regression line is superimposed on the scatterplot, its equation $y = 0.635x + 64.928$ is displayed, and the coefficient of determination $R^2 = 0.9888$ (in Excel's notation) is inserted on the scatterplot.

4. The output may be edited as previously with other charts for presentation purposes. For instance, activate the chart, and click on the rectangular box surrounding the equation; the border turns a darker grey. Use the **Decimal** tool to increase or decrease the number of decimal points. Edit the text by replacing x with "age" and y with "height." Move $R^2 = 0.9888$ from its location on the graph to a more convenient place. The final result with the regression line appears as Fig 2.14.

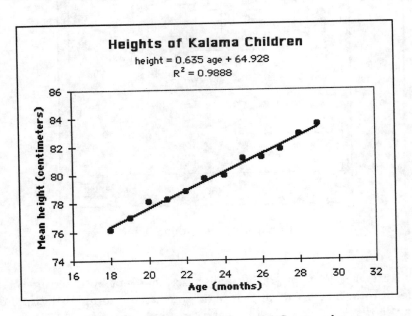

Figure 2.14: Regression Line and Scatterplot

Residuals

No discussion of regression is complete without an analysis of residuals, which provide evidence of how well the regression model fits. We defer discussion of this topic to Chapter 10 where the **Regression** tool will be introduced. However, in Fig 2.15 we show a scatterplot of residuals against age for the Kalama example. There does not appear to be any discernable pattern in the plot, indicating that a straight-line fit is appropriate.

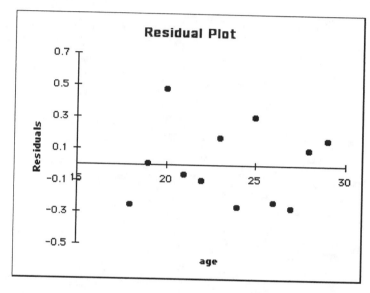

Figure 2.15: Residual Plot

Chapter 3

Producing Data

Excel provides tools for sampling from a specified population, for calculating probabilities associated with the standard models and their inverse cumulative distributions using the **Formula Palette**, and for simulating values from probability distributions using both the RAND() function and the **Random Number Generation** tool. In this chaper we consider both sampling without replacement and sampling with replacement (SRS) from a specified population.

3.1 Samples with Replacement

Using the Sampling Tool

Example 3.1. Simulate tossing a single die 10 times.

Solution. This is tantamount to finding a random sample of size 10 with replacement from $\{1, 2, 3, 4, 5, 6\}$. Following are the steps to obtain such a sample using the **Sampling tool**.

1. Enter the values $\{1, 2, 3, 4, 5, 6\}$ in A2:A7.

	A	B
1	Population	Sample
2	1	1
3	2	2
4	3	5
5	4	6
6	5	6
7	6	6
8		6
9		3
10		3
11		2

Figure 3.1: Sample with Replacement

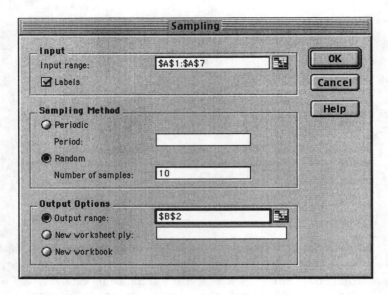

Figure 3.2: Sampling Dialog Box

2. From the Menu Bar choose **Tools − Data Analysis** and select **Sampling** from the dialog box. Click OK.

3. Complete the **Sampling** dialog box as shown in Fig 3.2 and click **OK**. A random sample of size 10 appears in cells B2:B11.

3.2   Simple Random Samples (SRS)

The Excel function RAND() picks a number uniformly on the interval (0, 1). We can repeatedly select random uniform (0, 1) numbers and assign them to members of a population. Then by sorting the random numbers, we obtain a random permutation of the population that provides an SRS of any desired size.

Example 3.2.   (Examples 3.6 and 3.7, pages 232–236 in the text.) A food company assesses the nutritional quality of a new "instant breakfast" product by feeding it to newly weaned male white rats and measuring their weight gain over a 28-day period. A control group of rats receives a standard diet for comparison. This nutrition experiment has a single factor (the diet) with two levels. The researchers use 30 rats for the experiment and so must divide them into two groups of 15. To do this in a completely unbiased fashion, they put the cage numbers of the 30 rats in a hat, mix them up, and draw 15. These rats form the experimental group and the remaining 15 make up the control group. Show how to carry out the randomization.

	A	B
1	Rats	Sample
2	1	0.75783563
3	2	0.25049354
4	3	0.30840101
5	4	0.01760057
6	5	0.06654875
7	6	0.78657626
8	7	0.17680760
9	8	0.63881337
10	9	0.99740651
11	10	0.74067272
12	11	0.35810086
13	12	0.11989300
14	13	0.68048251
15	14	0.23006162
16	15	0.64650203

Figure 3.3: Original Labels

Solution.

1. Give each rat a unique numerical label from the set $\{1, 2, 3, \ldots, 30\}$ and enter the values in cells A2:A31 of a workbook. Enter the label "Rats" in cell A1 and "Sample" in cell B1.

2. Enter $=$ RAND() in cell B2 and fill down to B31. The function RAND() selects a number uniformly in (0,1). Fig 3.3 shows a portion of the workbook.

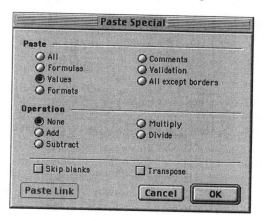

Figure 3.4: Paste Special Dialog Box

3. Select cells B2:B31 and from the Menu Bar choose **Edit** − **Copy**. Then, with B2:B31 **still selected**, choose **Edit** − **Paste Special** from the Menu Bar. (**Windows** users can click the **right mouse button** while **Macintosh** users should hold down the **Option** − **Command** keys and click to get the

Shortcut Menu box.) Select **Paste – Special**, and in the dialog box select the radio buttons for **Values** and **None** (Fig 3.4), which replaces the formulas in the cells of column B by the actual values they take.

4. Select cells A1:B31, from the Menu Bar choose **Data – Sort**, and in the Sort By drop-down list, click the arrow and select **Sample**. Also select the radio button for **Header Row** (Fig 3.5). Excel sorts the data in ascending order in column B and carries the order to column A, which gives a random permutation of column A.

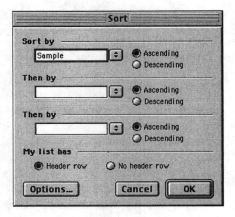

Figure 3.5: Sort Dialog Box

	A	B
1	Rats	Sample
2	4	0.01760057
3	30	0.03019514
4	17	0.06225799
5	5	0.06654875
6	12	0.11989300
7	24	0.14218590
8	27	0.17261367
9	7	0.17680760
10	26	0.18164762
11	14	0.23006162
12	2	0.25049354
13	29	0.29302707
14	3	0.30840101
15	20	0.34814598
16	21	0.35296414

Figure 3.6: Sorted Sample

5. Designate cells A2:A16 to label the rats in the control group, cells A17:A31 for the rats in the experimental group. A portion of the data appears in Fig 3.6, which may be compared with the original data in Fig 3.3.

3.3 Random Digits

Fig 3.7 is a table of random digits, a list of the digits $\{0, 1, \ldots, 9\}$ that has the following properties:

1. The digits in all positions in the list have the same chance of being any one of $\{0, 1, \ldots, 9\}$.

2. The digits in different positions are independent in the sense that the value of one has no influence on the value of any other.

You can imagine asking an assistant (or computer) to mix the digits $\{0, 1, \ldots, 9\}$ in a hat, draw one, then replace the digit drawn, mix again, draw a second digit, and so on. We did something like this in the previous section. In **Excel 97/98** this table of random digits dynamically changes when the F9 key on the keyboard is pressed. In **Excel 2000/2001** you need to hold down the Option key (Mac) or Control key (Windows) at the same time.

	A	B	C	D	E	F	G	H	I	J	K	L	M	N	O	P	Q	R	S	T
1																				
2									Table of Random Digits											
3									=INT(10*(RAND()))											
4	6	1	7	5	6	7	5	8	7	4	1	2	4	7	3	2	9	5	2	1
5	7	4	9	7	5	2	5	0	8	7	7	0	5	7	2	2	2	6	5	8
6	0	9	3	2	4	2	2	8	2	3	0	5	4	9	1	3	5	3	6	4
7	6	2	3	2	9	0	6	2	3	1	4	4	7	0	8	9	8	8	9	2
8	6	9	4	5	7	7	9	7	3	4	5	1	7	8	8	4	2	1	0	0
9	6	4	8	7	4	5	8	0	2	0	5	8	5	8	8	9	6	6	8	7
10	1	4	9	7	7	0	2	6	0	0	4	2	2	9	4	3	0	6	0	5
11	6	8	9	0	3	9	3	0	0	7	6	9	2	1	7	5	4	8	3	3
12	3	7	1	4	5	4	7	2	5	4	5	4	9	3	2	3	9	9	9	8
13	7	7	9	3	4	6	6	8	6	7	1	5	0	9	3	4	0	3	0	4
14	3	2	8	7	0	4	4	9	3	1	6	5	0	7	7	1	1	5	9	1
15	3	1	5	7	0	0	5	2	0	3	5	3	3	6	6	3	8	6	1	8
16	2	1	1	2	4	4	8	1	3	2	5	2	2	3	1	7	7	4	5	1
17	6	6	2	6	8	2	9	6	6	4	9	5	7	2	7	8	5	9	3	6
18	2	3	7	0	2	7	5	7	0	2	5	6	5	6	3	1	1	9	2	1
19	4	1	7	7	9	7	5	8	1	3	8	5	3	1	1	8	9	1	9	1
20	1	3	2	4	3	6	1	3	1	5	0	7	2	3	7	4	7	5	9	1
21	6	8	4	4	2	6	8	7	0	7	5	2	5	0	2	7	5	4	5	1
22	0	7	3	6	7	8	4	5	4	1	8	1	5	7	6	5	3	9	6	2
23	7	7	3	3	3	2	5	9	9	6	3	6	3	1	0	8	0	0	7	2
24	8	9	3	1	3	6	8	3	4	4	7	4	8	8	0	7	0	5	0	3
25	2	3	2	2	6	7	4	6	4	1	2	7	9	6	2	9	3	1	9	1
26	3	0	3	8	4	2	4	2	0	3	0	0	5	4	5	4	7	8	9	6
27	1	5	2	9	3	6	1	4	8	8	2	2	8	1	1	0	0	3	1	1
28	6	5	0	1	0	0	3	4	4	9	8	9	0	1	0	9	6	1	6	

Figure 3.7: Table of Random Digits

Using the RAND() Function

The Excel function `INT` truncates a real number to its integer value. For instance, the formula $= \mathtt{INT}(3.82)$ produces the value 3. By combining `INT` and `RAND()` as

$=$ INT(10*RAND()) we can produce random digits from $\{0, 1, \ldots, 9\}$.

Example 3.3. Produce a table of 500 random digits.

Solution

1. In cell A4 of a workbook, enter the formula $=$ INT(10*RAND()).

2. Select cell A4, click the fill handle in the lower right corner, and drag across to cell T4.

3. Select cells A4:T4, click the fill handle in the lower right corner, and drag down to cell T28.

The result is shown in Fig 3.7.

Chapter 4

Probability: The Study of Randomness

Probability models are used to describe and analyze real-world phenomena involving randomness. One way to develop an intuition for randomness is to observe random behavior. Computer simulations allow visual penetration into the concept of random variation.

4.1 Randomness

Simulating Bernoulli Random Variables

A real-world probability can only be estimated through the observation of data. Computer simulations are useful because they help develop insight into the meaning of random variation. Excel is well suited for simulation and provides both a RAND() function and a **Random Number Generation** tool for such purpose.

> **Example 4.1.** (Exercise 4.7, page 286 in the text.) The basketball player Shaquille O'Neal makes about half of his free throws over an entire season. Use Excel to simulate 100 free throws shot independently by a player who has probability 0.5 of making each shot. The technical term for independent trials with yes/no outcomes is Bernoulli trials. Our outcomes here are hit or miss.
> (a) What percent of the 100 shots did he hit in the simulation?
> (b) Examine the sequence of hits and misses. How long was the longest run of shots made? Of shots missed? (Sequences of random outcomes often show runs longer than our intuition thinks likely.)

Solution.   Again we will use the RAND() function to generate a sequence of 100 free throws and then invoke the **Chart Wizard** to dramatically display the results.

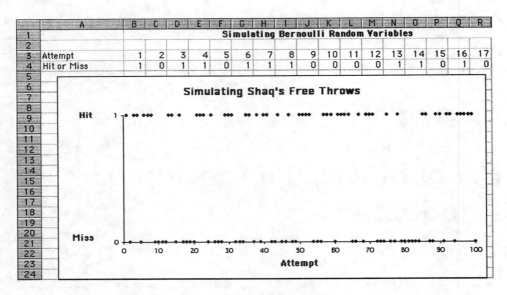

	A	B	C	D	E	F	G	H	I	J	K	L	M	N	O	P	Q	R
1									Simulating Bernoulli Random Variables									
3	Attempt	1	2	3	4	5	6	7	8	9	10	11	12	13	14	15	16	17
4	Hit or Miss	1	0	1	1	0	1	1	1	0	0	0	0	1	1	0	1	0

Figure 4.1: Simulating Shaq's Free Throws

1. Enter the labels "Attempt" in Cell A3 of a new workbook and "Hit" or "Miss" in cell A4. Enter the number "1" in cell B3 and fill to cell CW3 with successive integers $\{1, 2, \ldots, 100\}$. This can be achieved efficiently by selecting cell B3, then choosing **Edit − Fill − Series...** from the Menu Bar. Complete the **Series** dialog box with **Series in** Rows, **Type** Linear, **Step value** 1, and **Stop value** 100. Click OK.

2. Enter the = INT(2*RAND()) in cell B4. Select B4, click the fill handle, and drag to cell CW4 to generate 100 independent Bernoulli random variables.

In Fig 4.1 we show our workbook with the first 16 free throw simulations together with a graph of Shaq's hits or misses in 100 attempts. Notice how much more illuminating the graph is than the numerical sequence of hits and misses in showing random variation, including the presence of hot and cold streaks based on chance alone. The following steps explain how this chart is constructed.

Click the **ChartWizard** button.

Users of Excel 5/95

- In Step 1 enter the data range A3:CW4.
- In Step 2 click the **Scatter** chart type.
- In Step 3 select Format **1**.

- In Step 4 click the button for Data Series in **Rows**. Enter "1" for Use First Row for Category(X) Axis Labels and enter "1" for Use First 1 Column for Legend Text.

- In Step 5 select the radio button **No** for Add a legend?, and label the chart and X axis as shown in Fig 4.1.

- Click Finish.

Users of Excel 97/98/2000/2001

- In Step 1 click the **XY (Scatter)** Chart type and the first Chart sub-type (upper left on right side).

- In Step 2 on the **Data Range** tab, enter A3:CW4 for the range and check the radio button **Rows** for Series in:.

- In Step 3 on the **Titles** tab, enter the title and labels of the axes, on the **Axes** tab check both Category (X) axis and Category (Y) axis, on the **Gridlines** tab turn off all gridlines, on the **Legend** tab clear the legend, and finally on the **Data Labels** tab select the radio button **None**.

- In Step 4 embed the graph in the current workbook.

- Click Finish.

After the chart appears embedded on your workbook select it for editing and add the text "Hit" and "Miss" on the vertical axis.

Answers to Example 4.1 questions:
(a) By entering = SUM(B4:CW4)/100 in an empty cell, we find that the player hit 51% of his shots in the simulation.
(b) From Fig 4.1, we read that the longest run of hits is 5 and the longest run of misses is 6.
To simulate an additional 100 free throws, press the F9 key as indicated previously or resave the worksheet.

4.2   Probability Models

A probability model consists of a list of possible outcomes and a probability for each outcome (or interval of outcomes, in the case of continuous models). The probabilities are determined by the experiment that leads to the occurrence of one or more of the outcomes in the specified list.

Excel provides many distributions that may be constructed in a common fashion with the **Formula Palette** or the **Function Wizard**. The meaning of the required parameters is available online through Excel's help feature. Because of its prominence, the normal distribution was already discussed in Chapter 1. The binomial model will be considered in Chapter 5. Here we discuss some other distributions of particular interest in statistics.

Hypergeometric

HYPERGEOMDIST$(x, n, M, N,)$ provides probabilities for an experiment in which a simple random sample of size n is taken from a finite population of N individuals of which M are in a so-called "preferred category" called "success" or "1," while the remaining $N - M$ are deemed "failure" or "0." The function returns the probability of x successes in the sample of size n.

> **Example 4.2.** (Lotto 6/49) Suppose a box contains 49 balls in which one and only one ball is marked with an integer taken from $\{1, 2, \ldots, 49\}$. The balls are identical otherwise. Suppose that the balls numbered $\{1, 2, 3, 4, 5, 6\}$ are considered "successes." If an SRS of six balls is taken at random (without replacement), what is the probability that the sample contains k successes (for $k = 0, 1, 2, 3, 4, 5, 6$)?

	A	B	C
1		**Lotto 6/49 Probabilities**	
2			
3	k	P(X=k) value	P(K=k) formula
4	0	0.4359650	=HYPGEOMDIST(A4,6,6,49)
5	1	0.4130195	=HYPGEOMDIST(A5,6,6,49)
6	2	0.1323780	=HYPGEOMDIST(A6,6,6,49)
7	3	0.0176504	=HYPGEOMDIST(A7,6,6,49)
8	4	0.0009686	=HYPGEOMDIST(A8,6,6,49)
9	5	0.0000184	=HYPGEOMDIST(A9,6,6,49)
10	6	0.0000001	=HYPGEOMDIST(A10,6,6,49)

Figure 4.2: Calculating Lotto Probabilities

Solution. The answer is provided in Fig 4.2 where column B gives the probabilities and column C the corresponding Excel formulas.

Student t-Distribution

The Student t-distribution arises as the distribution of the *Studentized* score (similar to a standardized score)

$$t = \frac{\bar{x} - \mu}{s/\sqrt{n}}$$

where $\bar{x}, s$ are the sample mean and sample standard deviation from a sample of size n taken from a $N(\mu, \sigma)$ population. It is determined by a single parameter called the degrees of freedom ν. In the above ratio, ν takes the value $n - 1$.

Excel has an unusual definition of the c.d.f. and inverse c.d.f. for the t-distribution. The Excel function TDIST returns the tail of the distribution, that is, if $t(\nu)$ is a random variable with a t-distribution on ν d.f., then

$$\text{TDIST}(x, \nu, 1) = P[t(\nu) > x]$$

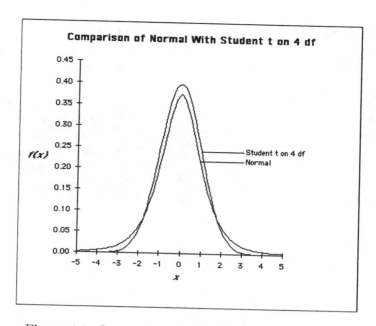

Figure 4.3: Comparing the Normal and the t Curves

and

$$\texttt{TDIST}(x, \nu, 2) = 2P[t(\nu) > x]$$

The argument x in $\texttt{TDIST}$ must be positive. Thus the c.d.f. takes a rather complicated expression using the logical $\texttt{IF}$ function:

$$P(t(\nu) \leq x) = \texttt{IF}(x < 0, \texttt{TDIST}(\texttt{ABS}(x), \nu, 1), 1 - \texttt{TDIST}(x, \nu, 1))$$

where $\texttt{ABS}(x) = |x|$, and similarly

$$P(|t(\nu)| \leq x) = 1 - \texttt{TDIST}(x, \nu, 2)$$

The inverse function $\texttt{TINV}$ is defined by

$$P[t(\nu) > \texttt{TINV}(\alpha, \nu)] = \frac{\alpha}{2}$$

so $\texttt{TINV}(\alpha, \nu)$ is the critical value for a two-sided significance test at level α of a normal mean (to be discussed in Chapter 7).

Exercise. Using $\texttt{NORMSDIST}$ and $\texttt{TDIST}$, graph on the same figure and to the same scale the densities of a $N(0, 1)$ and the Student t-distribution on 4 d.f. (Fig 4.3).

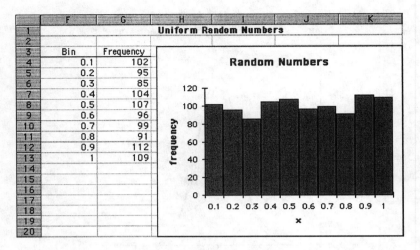

Figure 4.4: Simulating Uniform Random Variables

Uniform

A uniform random number is one whose values are spread out uniformly across the interval from 0 to 1. Its density curve has height 1 over the interval 0 to 1.

> **Example 4.3.** (Based on Example 4.17, page 310 in the text.) Let X be a uniform random number between 0 and 1. Use Excel to generate 1000 random uniform numbers, and from your simulations estimate the following probabilities and then compare them with the theoretical values.
>
> (a) $P(0.3 \leq X \leq 0.7)$
>
> (b) $P(X \leq 0.5)$
>
> (c) $P(X > 0.8)$

Solution.   Use RAND() to generate 1000 uniform random variables in a column and construct a histogram with bin intervals of width 0.10 beginning at 0 and ending at 1. Fig 4.4 shows the sample output from a workbook where this has been done. The frequencies shown are the number of times the random number generator produced a number X in the specified interval. The values listed under the heading *Bin* are the right endpoints of the intervals. We count the number of observations in the relevant intervals and divide by 1000 to convert to a probability.

(a) $P(0.3 \leq X \leq 0.7)= (104 + 107 + 96 + 99)/1000 = 0.406$.

(b) $P(X \leq 0.5)= (102 + 95 + 85 + 104 + 107)/1000 = 0.493$.

(c) $P(X > 0.8)= (112 + 109)/1000 = 0.221$.

The theoretical values are 0.400, 0.500, and 0.200, respectively.

Triangular—Adding Random Numbers

Example 4.4. (Based on Exercise 4.54, page 317 in the text.) Generate two random numbers between 0 and 1 and take Y to be their sum. Clearly the sum Y can take any number between 0 and 2. It is known that the idealized density curve of Y is a triangle. Use Excel to generate 1000 pairs of uniform random numbers, add them, and from your simulations estimate the following probabilities and compare them with the theoretical values.

(a) $P(0 \leq X \leq 0.5)$

(b) $P(0 \leq X \leq 1.0)$

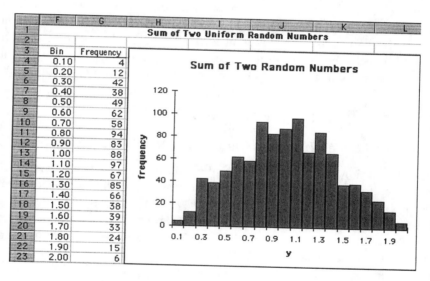

Figure 4.5: Simulating Triangular Random Variables

Solution. Again use RAND() to generate 1000 pairs of uniform random variables in two columns, add the columns and construct a histogram with bin intervals of width 0.10 beginning at 0 and ending at 2. Fig 4.5 shows the sample output from a workbook where this has been done. The frequencies shown are the number of times the random number generator produced a number in the specified interval.

(a) $P(0 \leq X \leq 0.5) = (4 + 12 + 42 + 38 + 49)/1000) = 0.145$.

(b) $P(0 \leq X \leq 1.0) = (4+12+42+38+49+62+58+94+93+88)/1000 = 0.540$.

The corresponding theoretical values are 0.125 and 0.500, respectively.

4.3   Random Variables

A random variable is completely prescribed by its probability distribution. This distribution has a long-run frequency interpretation associated with the **Law of Large Numbers**.

> Draw independent observations at random from any population with finite mean μ. Decide how accurately you would like to estimate μ. As the number of observations drawn increases, the mean $\bar{x}$ of the observed values eventually approaches the mean μ of the population as closely as you specified and then stays that close.

Applying this phenomenon to a discrete random variable X, suppose that

$$P(X = 3) = 0.5$$

In repeated trials, consider the proportion $\hat{p}$ of times that X takes the value 3. The random variable $\hat{p}$ is the sample mean of a sequence of random variables taken from a Bernoulli population with probability of success $= 0.5$ and whose population mean is therefore also 0.5. The Law of Large Numbers then asserts that in some sense (made precise by the theory of probability) $\hat{p}$ approaches 0.5 as the number of trials increases, which gives a relative frequency interpretation of probability.

Simulating Random Variables

The Law of Large Numbers Using the RAND() Function

The Excel function RAND() picks a number uniformly on the interval (0,1). To generate a uniform random variable on (a, b) use $a + \text{RAND()} * (b - a)$. Using the inverse probability function $h(a) = \inf\{x : F(x) \geq a\}$, we can then generate other distributions. Thus NORMINV(RAND(), Mean, StDev) returns a random normal with mean given by Mean (either a numerical value or a named reference to a numerical value) and standard deviation by StDev.

By examining the list of functions available (clicking the **Paste Function** button f_x on the Standard Toolbar), you can determine which distributions Excel can simulate this way and how to describe the required parameters.

> **Example 4.5.**   Using the RAND() function, simulate 1000 independent Bernoulli trials based on tossing a fair coin, calculate the cumulative proportion of heads $\hat{p}$ after each trial, and construct a graph that demonstrates the law of large numbers in action. Also show on the same graph a horizontal line at the height 0.5

Solution.   The RAND() function produces a number uniformly distributed on the interval (0,1). This can be converted into integers taking the values 0 or 1 with equal probability if this uniform random number is multiplied by 2 and then the integer part is taken. The Excel formula for these operations is $= \text{INT(2*RAND())}$.

1. Enter the formula = `INT(2*RAND())` in cell A5 of a new workbook and copy this formula down to cell A1003 by selecting cell A4, then clicking the fill handle and dragging to cell A1003 to generate 1000 tosses of a fair coin (0 representing tails and 1 representing heads).

2. Enter the value "0" in cell B3 followed by the formula = A4+B3 in cell B4. Copy the formula in cell A4 down to cell A1003. Column B tracks the cumulative number of heads.

3. Enter the number 1 in cell C4 and fill to cell C1003 with successive integers $\{1, 2, \ldots, 1000\}$. This can be achieved efficiently by selecting cell C4 and then choosing **Edit − Fill − Series** from the Menu Bar. Complete the **Series** dialog box with Series in **Columns**, Type **Linear**, and **Step Value** 1, **Stop Value** 1000. Click OK. Column C will label the 1000 tosses.

4. Fill cells D4 to D1003 with the value 0.5. This will represent the horizontal line at height 0.5 on the graph.

5. Enter the formula = B4/C4 in cell E4 and copy to cell E1003.

Fig 4.6 shows part of the workbook with the required formulas.

	A	B	C	D	E
1	Formulas Behind Simulation of 1000 Tosses				
2					
3		0			
4	=INT(2*RAND())	=A4+B3	1	0.5	=B4/C4
5	=INT(2*RAND())	=A5+B4	2	0.5	=B5/C5
6	=INT(2*RAND())	=A6+B5	3	0.5	=B6/C6

Figure 4.6: Simulating 1000 Tosses of a Fair Coin

We next construct a graph displaying the same results. Click the **ChartWizard** button.

Users of Excel 5/95

- In Step 1 enter the data range C4:E1003.

- In Step 2 click the **Line** chart type.

- In Step 3 select Format **1**.

- In Step 4 click the button for Data Series in **Columns**. Enter "1" for Use First 1 Column for Category(X) axis labels and enter "0" for Use First 0 Column for Legend Text.

- In Step 5 click the radio button **No** for Add a legend? and label the chart and X axis as shown in Fig 4.7. Click Finish.

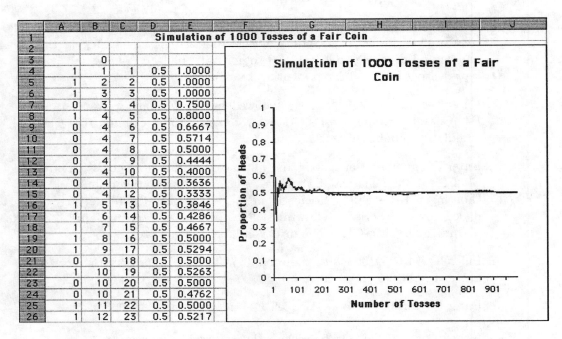

	A	B	C	D	E
1			Simulation of 1000 Tosses of a Fair Coin		
2					
3		0			
4	1	1	1	0.5	1.0000
5	1	2	2	0.5	1.0000
6	1	3	3	0.5	1.0000
7	0	3	4	0.5	0.7500
8	1	4	5	0.5	0.8000
9	0	4	6	0.5	0.6667
10	0	4	7	0.5	0.5714
11	0	4	8	0.5	0.5000
12	0	4	9	0.5	0.4444
13	0	4	10	0.5	0.4000
14	0	4	11	0.5	0.3636
15	0	4	12	0.5	0.3333
16	1	5	13	0.5	0.3846
17	1	6	14	0.5	0.4286
18	1	7	15	0.5	0.4667
19	1	8	16	0.5	0.5000
20	1	9	17	0.5	0.5294
21	0	9	18	0.5	0.5000
22	1	10	19	0.5	0.5263
23	0	10	20	0.5	0.5000
24	0	10	21	0.5	0.4762
25	1	11	22	0.5	0.5000
26	1	12	23	0.5	0.5217

Figure 4.7: Law of Large Numbers

Users of Excel 97/98/2000/2001

- In Step 1 click the **Line** chart type and the second Chart sub-type **Line**.

- In Step 2 on the **Data Range** tab enter D4:E1003 for the Data range check the radio button **Series in: Columns**.

- In Step 3 on the **Titles** tab, enter the title and labels of the axes, on the **Axes** tab check radio button **Automatic** for Category (X) axis and check the Value (Y) axis box, on the **Gridlines** tab turn off all gridlines, on the **Legend** tab clear the legend, and finally on the **Data Labels** tab select the radio button **None**.

- In Step 4 embed the graph in the current workbook. Click Finish.

Format the X and Y axes as shown in Fig 4.7, for instance by changing the number of categories between tick marks and reorienting the X axis labels.

Fig 4.7 shows a segment of the completed workbook with the embedded graph. As previously you can reevaluate all functions and the graph will dynamically change. From column A you can see the random sequence of heads and tails generated, while column E exhibits the proportions of heads. These are quite variable at first but then settle down, appearing to approach the value 0.5 (shown by the horizontal line). This behavior is known in statistics as a law of large numbers, commonly referred to as the "law of averages."

Simulating Random Variables Using Random Number Generation

In addition to the function RAND(), Excel has a **Random Number Generation** tool built into the **Analysis ToolPak** that provides an alternative and more systematic approach to simulation.

The **Random Number Generation** tool creates columns of random numbers, as specified by the user, from any of six probability models (uniform, normal, Bernoulli, binomial, Poisson, discrete) as well as having an option for patterned that creates not random data but rather data according to a specified pattern.

All options are invoked from a common dialog box (as in Fig 4.8 for discrete) following the choice **Tools – Data Analysis – Random Number Generation** from the Menu Bar. Select the distribution of interest using the drop-down arrow and the Parameters sub-box will automatically change, prompting input of parameters.

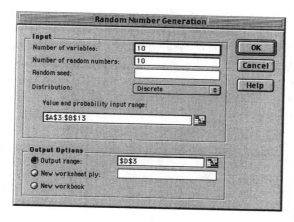

Figure 4.8: Random Number Generation Tool—Discrete

Number of Variables. Enter the number of columns of random variables. The default is all columns.

Number of Random Numbers. Enter the number of rows (cases) of random variables.

Distributions. Use the drop-down arrow to open a list of choices with requested parameters.

> **Uniform.** Upper and lower limits
>
> **Normal.** μ, σ
>
> **Bernoulli.** $p =$ probability of success; Excel unfortunately refers to this as a p Value.

Binomial. p, n

Poisson. λ

Discrete.   Specify the possible values and their corresponding probabilities. Before using this option enter the values and probabilities in adjacent columns in the workbook.

Patterned.   This option creates data according to a prescribed pattern of values repeated in specified steps. This is useful if a linear array of data needs to be coded using another variable.

Example 4.6.   To generate 100 tosses of a pair of fair dice enter $\{2, 3, \ldots, 11\}$ into cells A3:A13 and enter

$$\{1/36, 2/36, \ldots, 6/36, 5/36, \ldots, 2/36, 1/36\}$$

into cells B3:B12 (Fig 4.9). Excel may interpret the value 1/36 as a date Jan 1936. If this happens then format the cells by choosing **Format − Cells** from the Menu Bar and selecting **Number**. Then choose **Tools − Data Analysis − Random Number Generation** from the Menu Bar and complete as in Fig 4.8. The output will appear in cells D1:M10. Since these numbers are random your output will of course be different.

	A	B	C	D	E	F	G	H	I	J	K	L	M
1		Simulation of a Pair of Fair Dice											
2	k	P(X=k)											
3	2	0.0277778		2	8	6	8	4	7	7	8	6	5
4	3	0.0555556		6	9	5	6	8	9	12	9	7	7
5	4	0.0833333		8	5	8	8	12	6	7	8	9	5
6	5	0.1111111		2	9	3	8	5	7	7	6	5	8
7	6	0.1388889		8	6	11	10	6	8	4	8	7	8
8	7	0.1666667		10	6	9	8	2	6	4	7	6	4
9	8	0.1388889		9	3	4	5	10	7	7	6	4	7
10	9	0.1111111		6	6	4	11	9	7	7	2	11	4
11	10	0.0833333		6	9	8	9	5	8	3	12	3	4
12	11	0.0555556		12	8	3	5	8	4	10	7	10	7
13	12	0.0277778											

Figure 4.9: Simulating a Pair of Fair Dice

Chapter 5

From Probability to Inference

The probability distribution of a statistic obtained from an experiment is called its sampling distribution. An important class of statistics arises when the observations are counts of some variable. This leads to the binomial model for sample counts and sample proportions. These sampling distributions can be approximated by normal curves, and they directly demonstrate several important results about sample means $\bar{x}$ in general:

1. $\bar{x}$ is an unbiased estimate of the population mean μ.

2. The standard deviation of $\bar{x}$ is equal to $\frac{\sigma}{\sqrt{n}}$ where n is the sample size and σ the population standard deviation.

3. The sampling distribution of $\bar{x}$ is approximately $N(\mu, \frac{\sigma}{\sqrt{n}})$.

5.1 Sampling Distributions for Counts and Proportions

The Binomial Distribution

A binomial distribution is associated with an experiment comprising n independent trials each of which has the same success probability p. The random variable X counts the number of successes.

It is known that

$$P(X = k) = \binom{n}{k} p^k (1-p)^{n-k} \qquad k = 0, 1, 2, \ldots, n$$

$$\text{mean} = \mu_X = np$$

and

$$\text{standard deviation} = \sigma_X = \sqrt{np(1-p)}$$

The corresponding Excel function is `BINOMDIST`$(k, n, p,$ cumulative$)$. If the parameter cumulative is set to "false," Excel returns the probabilities $P(X = k)$, while if it is set to "true," Excel returns the cumulative probabilities $P(X \le k)$.

Example 5.1.   Construct a binomial table for $n = 15$ and $p = 0.03$, including both individual and cumulative probabilities.

Solution

1. Enter the label k in cell A1 and the label $P(X = k)$ in cell B1 of a new workbook. In A2:A17 enter the values $\{0, 1, 2, \ldots, 15\}$.

2. Activate cell B2. Using either the **Formula Palette** or the **Function Wizard**, construct the binomial function by selecting **Statistical** for Function Category and `BINOMDIST` for Function Name.

3. Input the following into the dialog box

    number_s   Enter the cell address A2
    trials   Enter the value 15
    probability_s   Enter the value 0.3
    cumulative   Enter the value 0

    to create the formula `BINOMDIST(A2,15,0.3,0)`. Click Finish or OK.

4. Activate cell B2, click the fill handle in the lower right corner, and drag to cell B17 to fill the column with individual binomial probabilities (Fig 5.1).

	A	B	C
1	k	P(X=k)	P(X<=k)
2	0	0.00475	0.00475
3	1	0.03052	0.03527
4	2	0.09156	0.12683
5	3	0.17004	0.29687
6	4	0.21862	0.51549
7	5	0.20613	0.72162
8	6	0.14724	0.86886
9	7	0.08113	0.94999
10	8	0.03477	0.98476
11	9	0.01159	0.99635
12	10	0.00298	0.99933
13	11	0.00058	0.99991
14	12	0.00008	0.99999
15	13	0.00001	1.00000
16	14	0.00000	1.00000
17	15	0.00000	1.00000

Figure 5.1: Binomial Probabilities

5. Next label cell C1 as $P(X <= k)$ and repeat Steps 2, 3, and 4. Activate C2 instead of B2 in Steps 2 and 4 and enter the value 1 for the cumulative distribution in Step 3.

The resulting table of individual and cumulative binomial probabilities appears in Fig 5.1.

Binomial Distribution Chart

We can quickly construct a histogram using the **ChartWizard** displaying the binomial probabilities just calculated. As the procedure is identical to earlier constructions of charts, we omit the details. This histogram appears in Fig 5.2.

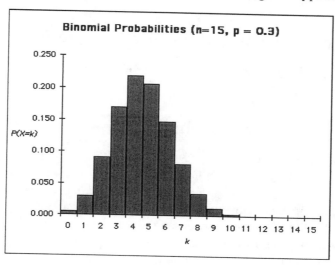

Figure 5.2: Binomial Histogram

Inverse Cumulative Binomial

`CRITBINOM`(trials, probability_s, alpha) returns the smallest x for which the binomial cumulative distribution function (c.d.f.) is greater than or equal to alpha; that is, if $B(x)$ represents the binomial c.d.f. on n trials and success probability p, then $=$ `CRITBINOM`(n, p, α) returns

$$B^{-1}(\alpha) = \inf\{x : B(x) \geq \alpha\}, \quad 0 < \alpha \leq 1$$

For $\alpha = 0$, this definition gives $-\infty$ and Excel gives the error message `#NUM!`.

Inverse probabilities are useful for finding P-values and in **simulation** because from the definition, if U is uniform $(0, 1)$ and $F(x)$ is an arbitrary c.d.f. with inverse defined by

$$F^{-1}(\alpha) = \inf\{x : F(x) \geq \alpha\}$$

then

$$X = F^{-1}(U)$$

has the specified distribution $F(x)$. Thus, for instance

$$= \texttt{CRITBINOM}(u, p, U)$$

is a binomial random variable on n trials and success probability p, and

$$= \texttt{NORMINV}(U, \mu, \sigma)$$

is a $N(\mu, \sigma)$ random variable.

5.2   Sampling Distribution of a Sample Mean

Simulation followed by a histogram of the results provides an insightful view of the *central limit theorem*.

Simulating the Central Limit Theorem

Example 5.2.   (Exercise 5.36, page 404 in the text.)   A roulette wheel has 38 slots – 18 are black, 18 are red, and 2 are green. When the wheel is spun, a ball is equally likely to come to rest in any of the slots. Gamblers can place a number of different bets in roulette. One of the simplest wagers chooses red or black. A bet of one dollar on red will pay off an additional dollar if the ball lands in a red slot. Otherwise the player loses his dollar. When a gambler bets on red or black, the two green slots belong to the house. A gambler's winnings on a \$1 bet are either \$1 or −\$1.

(a) Simulate a gambler's winnings after 50 bets and compare the gambler's mean winnings per bet with the theoretical results.

(b) Compare the results with the normal approximation.

Solution. The number of wins after 50 bets X is a binomial $B(50, 10/38)$ random variable with

$$\text{mean} \qquad \mu_X = 50 \left(\frac{18}{38}\right) = 23.684$$

$$\text{standard deviation} \qquad \sigma_X = \sqrt{50 \left(\frac{18}{38}\right)\left(\frac{20}{38}\right)} = 3.5306$$

The proportion of wins after 50 bets is $\hat{p} = X/50$ with

$$\text{mean} \qquad \mu_{\hat{p}} = \frac{18}{38} = 0.4737$$

$$\text{standard deviation} \qquad \sigma_{\hat{p}} = \sqrt{\left(\frac{18}{38}\right)\left(\frac{20}{38}\right) \Big/ 50} = 0.0706$$

The gambler either wins \$1 or loses \$1. His average winnings per game, denoted by $\bar{w}$, are therefore $1\times$ the proportion of times that $X = 1$ minus $1\times$ the proportion of times that $X = -1$, that is $\bar{w} = \hat{p}(1) + (1 - \hat{p})(-1) = 2\hat{p} - 1$. By the rules for means and standard deviations

$$\text{mean} \qquad \mu_{\bar{w}} = 2\mu_{\hat{p}} - 1 = -0.0527$$
$$\text{standard deviation} \qquad \sigma_{\bar{w}} = 2\sigma_{\hat{p}} = 0.14123$$

By the **Central Limit Theorem** $\hat{p}$ is approximately normal, and therefore

$$\bar{w} \text{ is approximately } N(-0.0527, 0.14123)$$

We can simulate a binomial random variable X, convert it first to $\hat{p} = \frac{X}{n}$ and then to $\bar{w} = 2\hat{p} - 1$, after which we construct a histogram of the simulation results.

The following steps, referring to Fig 5.3, show how to develop a workbook to simulate 500 replications of 50 games. We will use the RAND function, which *links* the output to a histogram.

	A	B	C	D	E
1			Demonstrating the Central Limit Theorem by Simulation		
2					
3	true mean =	-0.0527	simulated mean =	-0.0556	
4	true st_dev =	0.14123	simulated st_dev =	0.14120	
5					
6	Formula entered in column B		=2*CRITBINOM(50,18/38,RAND())/50 - 1		
7	Simulation	Average per Game			
8	1	0.040			
9	2	-0.200	Bin Formulas	Bin	Freq.
10	3	-0.040	=-0.0527-3.5*0.14123	-0.55	0
11	4	-0.080	=-0.0527-3*0.14123	-0.48	1
12	5	-0.120	=-0.0527-2.5*0.14123	-0.41	0
13	6	-0.280	=-0.0527-2*0.14123	-0.34	9
14	7	0.120	=-0.0527-1.5*0.14123	-0.26	26
15	8	0.000	=-0.0527-0.14123	-0.19	57
16	9	-0.040	=-0.0527-0.5*0.14123	-0.12	54
17	10	-0.160	=-0.0527	-0.05	86
18	11	-0.080	=-0.0527+0.5*0.14123	0.02	119
19	12	-0.160	=-0.0527+0.14123	0.09	85
20	13	0.040	=-0.0527+1.5*0.14123	0.16	27
21	14	-0.120	=-0.0527+2*0.14123	0.23	26
22	15	0.160	=-0.0527+2.5*0.14123	0.30	7
23	16	0.040	=-0.0527+3*0.14123	0.37	1
24	17	-0.040	=-0.0527+3.5*0.14123	0.44	2
					500

Figure 5.3: Simulating the Central Limit Theorem

Bin intervals for the histogram will be located at multiples of the standard deviation from the mean

1. Prepare a new workbook by entering "Simulating the Central Limit Theorem" in cell A1 and centering the heading across A1:E1. Enter "Simulation" in A7 and "Average per Game" in B7.

2. Enter the values $1, 2, \ldots, 500$ in cells A8:A507 as follows: Enter "1" in A8. Select A8 and choose **Edit – Fill – Series...** from the Menu Bar. In

the **Series** dialog box, check Series in **Columns** and Type **Linear**. Clear the **Trend** box and type "1" and "500" for the **Step** and **Stop** values, respectively.

3. Now simulate 50 games. In cell B8 enter $= 2*\text{CRITBINOM}(50,18/38,\text{RAND}())/50-1$ to generate the random variable $\bar{w}$. We have shown the formula on line 6 beginning in column C. Select B8, click the fill handle at the lower right corner of B8 and drag down to cell B507. Cells B8:B507 are now filled with 500 replications of the gambler's average net gain per game after each simulated 50 games.

4. Next we prepare the simulations for output into a histogram. Enter the labels "Bin" in C8 and "Freq." in D8. The bin endpoints for the histogram appear in cells D9:D22 and the corresponding formulas behind the values are shown in cells C9:C23. These are based on theoretical true mean and standard deviation and the bin endpoints are expressed in simple multiples of the standard deviation from the mean, Enter $= -0.0527 - 3.5 * 0.14123$ in D9, $= -0.0527 - 3.0 * 0.14123$ in D10, and so on. Refer to Fig 5.3 where we have shown in cells C9:C23 the formulas to be entered in D9:D23. Note that you do not require a column C in your own worksheet.

5. Select E9:E23. Then type $= \text{FREQUENCY}$(B8:B507,D9:D23) in the entry area of the **Formula Bar**. Hold down the **Shift and Control** keys (either **Macintosh** or **Windows**) and press enter/return to **array-enter** the formula. The formula will appear **surrounded by braces** { } in the **Formula Bar**, and the bin frequencies will appear in cells E9:E23. Select cells D9:D23 and then complete the sequence of steps in the **Chart Wizard** as discussed previously. The resulting histogram appears in Fig 5.4.

6. Note that we have also located the sample mean $\bar{w}$ and sample standard deviation $s_{\bar{w}}$ on the worksheet for comparison with the true values. These will change whenever the worksheet is re-evaluated. The formula behind cell D3 is =AVERAGE(B8:B507) and behind cell D4 it is $= \text{SQRT}((1-D3*D3)/50)$.

Excel Output

The sample mean and sample standard deviation appear in D3:D4, and the population mean and standard deviation appear in B3:B4 for comparison purposes. For the simulation shown

$$\bar{w} = -0.0536 \qquad \mu_{\bar{w}} = -0.0527$$
$$s_{\bar{w}} = 0.14120 \qquad \sigma_{\bar{w}} = 0.14123$$

The table of frequency counts appears in E9:E23 with the corresponding histogram in Fig 5.4. The histogram appears normal shaped with no unusual features.

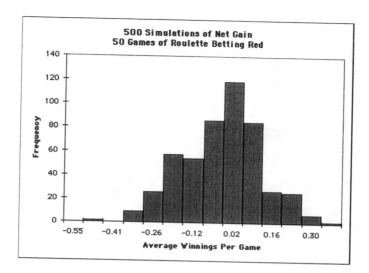

Figure 5.4: Histogram of 500 Simulations of 50 Roulette Games

Recalling that the bin entries are the right endpoints of the bin interval, we can determine the proportion of counts within 1, 2, and 3 standard deviation units of the mean. The simulation results are alarmingly good.

	Actual	Theoretical
Within 1 σ	.688	.683
Within 2 σ	.960	.954
Within 3 σ	.994	.997

Chapter 6

Introduction to Inference

This chapter discusses procedures for finding confidence intervals and carrying out significance tests for the mean of a population. The methods require use of the normal distribution and hence are applicable only when the underlying population may be assumed to be approximately normal or when the sample size is so large that the normal approximation given by the central limit theorem may be invoked.

6.1 Estimating with Confidence

The data $\{x_1, \ldots, x_n\}$ are assumed to come from a $N(\mu, \sigma)$ population with mean μ and a known standard deviation σ. A level C confidence interval for the mean μ is given by

$$\bar{x} \pm z^* \frac{\sigma}{\sqrt{n}}$$

where $\bar{x} = \frac{1}{n} \sum_{i=1}^{n} x_i$ is the sample mean, n is the sample size, and z^* is the value such that the area between $-z^*$ and z^* under a standard normal curve equals C.

The Excel function required is $= \text{CONFIDENCE}(\alpha, \sigma, n)$ which calculates the margin of error (half width of interval) associated with a $C = 1 - \alpha$ level confidence interval for the mean of a normal distribution with standard deviation σ based on a sample of size n, that is $z^* \frac{\sigma}{\sqrt{n}}$

It is also possible to calculate each component in the confidence interval directly. This more complicated approach is useful if intermediate results such as margin of error are needed. The formula for the normal critical value z^* is $\text{NORMSINV}(p)$, which returns the inverse of the standard normal cumulative distribution function $\Phi^{-1}(p)$, where $0 < p < 1$. In the first example below a confidence interval is constructed both ways.

> **Example 6.1.** (Example 6.2, page 422 in the text.) Tim Kelley has been weighing himself once a week for several years. Last month his four measurements (in pounds) were

$$190.5 \qquad 189.0 \qquad 195.5 \qquad 187.0$$

	A	B	C	D
1	Confidence Interval for a Normal Mean Using Z			
2				
3		values	formulas	
4	User Input			
5	sigma	3		Data
6	conf	0.90		190.5
7	Summary Statistics			189.0
8	n	4	=COUNT(Data)	195.5
9	xbar	190.5	=AVERAGE(Data)	187.0
10	Calculations			
11	SE	1.500	=sigma/SQRT(n)	
12	z	1.64	=NORMSINV(0.5+conf/2)	
13	ME	2.47	=z*SE	
14	Excel ME	2.47	=CONFIDENCE(1-conf,sigma,n)	
15	Confidence Limits			
16	lower	188.03	=xbar-ME	
17	upper	192.97	=xbar+ME	

Figure 6.1: Confidence Interval for a Normal Mean

Give a 95% confidence interval for his mean weight last month.

Solution. In Fig 6.1 the Excel formulas required are given in column C. These are entered into the adjacent cells in column B to create the workbook template to solve this problem. The Excel output is in column B. The user inputs required are the standard deviation and the confidence level. If the data have already been summarized, then you can enter the values for the sample size n and the average in cells B8:B9, respectively. Otherwise Excel will read the data and calculate n and $\bar{x}$. The data can be located in a convenient place on the same sheet or it can be located on another sheet. The latter is particularly useful for large data sets. In this example, with only four data points, we have recorded them on the same sheet as the calculations.

The following steps describe how to construct the workbook.

1. Enter the labels as shown in column A.

2. **Name** the cell ranges to be used. Select cells A5:B6, A8:B9, A11:B13 and A16:B17. To select **noncontiguous blocks** of cells, make the first selection A5:B6, then hold down the **Control (Windows)** or **Command (Macintosh)** key while selecting the other ranges. From the Menu Bar choose **Name – Create**, select **Left Column**, and then click OK. Next, select D5:D9, and from the Menu Bar choose **Name – Create**, select **Top Row**, and then click OK to name the data range.

3. Enter the formulas shown in Fig 6.1 into columns B8:B9, B11:B14, and B16:B17, and enter the data in D5:D9. Since you have named the data range you can refer to the cells D5:D9 as "Data," for instance as in the formula = COUNT(Data). Otherwise, you would type = COUNT(D5:D9) giving the actual

locations. These formulas are sufficiently simple that you can enter them by hand rather than use the Formula Palette or the Function Wizard.

The only input needed once the workbook has been constructed are the population standard deviation (sigma) and the confidence level. Type "3" and "0.90" into cells B5 and B6, respectively. The results are immediately recorded in cells B16:B17, showing a lower confidence limit 188.03 and an upper confidence limit 192.97 for the population mean μ.

Explanation

The formula $=$ COUNT(Data) gives the sample size by counting the number of cells named by the variable Data. You could also type the integer "4" instead. Likewise, $=$ AVERAGE(Data) is the Excel formula for the sample mean $\bar{x}$. For comparison we have also provided the corresponding formula $=$ CONFIDENCE(α, σ, n) in cell C14.

How Confidence Intervals Behave

A confidence interval is a random interval that has a specified probability of containing an unknown parameter. Thus, a 90% confidence interval for a population mean has probability 0.90 of containing the mean. So, in repeated confidence intervals, in the long run approximately 90% of these confidence intervals would contain the population mean.

> **Example 6.2.** Take 100 SRS of size 3 from an $N(3.0, 0.2)$ population and construct a 90% confidence interval for the mean. Count how many times the confidence interval contains the mean 3.0.

Solution

1. Following the instructions given in Example 4.6 for simulating samples from a specified distribution, choose **Tools – Data Analysis – Random Number Generation** from the Menu Bar, complete a box like the one shown in Fig 4.8, but for normal not discrete random numbers, with "3" for the **Number of Variables**, "100" for the **Number of Random Numbers**, "3.0" for the **Mean**, "0.2" for the **Standard Deviation**, and choose a convenient range for the output. We have selected the range A8:C107.

2. In cell E8 enter
   $=$ AVERAGE(A8:C8) $-$ NORMSINV(0.5+0.9/2)*0.2/SQRT(3)
   In cell F8 enter
   $=$ AVERAGE(A8:C8) $+$ NORMSINV(0.5+0.9/2)*0.2/SQRT(3)

3. Select cells E8:F8, click the fill handle and drag the contents to F107. The cells in column F will contain the value 1 if the confidence interval for the data in the corresponding row contains the true value 3.0, while the cells will contain 0 otherwise.

4. Count the number of times 1 appears by entering $=$ SUM(G8:G107) in an empty cell (H8, for example).

Fig 6.2 shows a portion of a workbook with the simulation for which 92 times out of 100 the true mean was within the 90% confidence limits.

	A	B	C	D	E	F	G	H
1	\multicolumn{8}{c}{**Behavior of Repeated Confidence Intervals**}
2								
3	lower	= AVERAGE(A8:C8) −NORMSINV(0.5+0.90/2)*0.2/SQRT(3)						
4	upper	= AVERAGE(A8:C8) −NORMSINV(0.5+0.9/2)*0.2/SQRT(3)						
5	G8	=IF(AND(E8<3, 3<F8), 1,0)						
6								
7					lower	upper		
8	3.1772	2.7218	3.3097		2.880	3.259	1	92
9	3.0863	3.0417	2.8220		2.793	3.173	1	
10	2.8207	2.8480	2.9353		2.678	3.058	1	
11	2.9131	3.2380	3.0292		2.870	3.250	1	
12	3.0904	3.1497	2.9295		2.867	3.246	1	
13	2.8767	3.0868	3.2555		2.883	3.263	1	
14	2.8937	2.7254	2.9995		2.683	3.063	1	
15	3.1976	3.0303	2.8750		2.844	3.224	1	
16	2.8378	2.9206	2.7565		2.648	3.028	1	
17	2.7972	2.9133	3.1956		2.779	3.159	1	
18	2.5773	3.2215	3.0810		2.770	3.150	1	
19	3.1855	2.6901	3.0221		2.776	3.156	1	
20	3.1730	3.1092	3.0020		2.905	3.285	1	

Figure 6.2: Repeated Confidence Intervals

6.2   Tests of Significance

Significance tests are used to judge whether a specified (null) hypothesis is consistent with a data set.

We create a workbook for testing the null hypothesis $H_0 : \mu = \mu_0$ for a specified null value μ_0 against one-sided or two-sided alternatives. The data $\{x_1, x_2, \ldots, x_n\}$ are assumed to come from an $N(\mu, \sigma)$ population where σ is known. The same procedure can also be used to carry out a large sample test. In the workbook in Fig 6.3, the user can either test at a specified level of significance or determine a P-value.

The user inputs are the sample size, sample mean, standard deviation (which may be input as values, as formulas, or as named references depending on the context), null hypothesis, and level of significance.

> **Example 6.3.**   (Example 6.16, page 450 in the text.) Bottles of a popular cola drink are supposed to contain 300 ml of cola. There is some variation from bottle to bottle because the filling machinery is not precise. The distribution of the contents is normal with standard deviation $\sigma = 3$ ml. A student who suspects that the bottle is underfilling measures the contents of six bottles. The results are 299.4, 297.7, 310.0, 298.9, 300.2, 297.0. Is this convincing evidence that the mean content of cola bottles is less than the advertised 300 ml?

	A	B	C	D
1		Z Test for a Normal Mean – Values and Formulas		
2				
3		values	formulas	
4	User Input			
5	sigma	3.0		Data
6	null	300.0		299.4
7	alpha	0.05		297.7
8				301.0
9	Summary Statistics			298.9
10	n	6	=COUNT(Data)	300.2
11	xbar	299.03	=AVERAGE(Data)	297.0
12	Calculations			
13	SE	1.225	=sigma/SQRT(n)	
14	z	-0.789	=(xbar-Null)/SE	
15	Lower Test			
16	lower_z	-1.645	=NORMSINV(alpha)	
17	Decision	Do Not Reject H0	=IF(z<lower_z,"Reject H0","Do Not Reject H0")	
18	Pvalue	0.215	=NORMSDIST(z)	
19	Upper Test			
20	upper_z		=-NORMSINV(alpha)	
21	Decision		=IF(z>upper_z,"Reject H0","Do Not Reject H0")	
22	Pvalue		=1-NORMSDIST(z)	
23	Two-Sided Test			
24	two_z		=ABS(NORMSINV(alpha/2))	
25	Decision		=IF(ABS(z)>two_z,"Reject H0","Do Not Reject H0")	
26	Pvalue		=2*(1-NORMSDIST(ABS(z)))	

Figure 6.3: Significance Test for a Normal Mean

Solution. The workbook template in Fig 6.3 shows all formulas required in column C for any type of alternative: lower, upper, and two-sided tests, respectively. These go into the adjacent cells of column B. Then enter the values sigma = 3, alpha = 0.05, and null hypothesis = 300.0. Here the alternative is $H_a : \mu < \mu_0$ and only the values for the lower test are therefore shown in column B. The critical value at the 5% level of significance is $-z^* = -1.645$ in cell B16 and since the computed z in B14 is -0.789 we do not reject H_0. Note that the P-value 0.215 is also given in B18.

Explanation

We encountered the function NORMSINV previously. The formula = NORMSDIST returns the cumulative normal distribution function $\Phi(z)$. For a one-sided lower test, the P-value is the area to the left of the computed z score $\frac{\bar{x}-\mu_0}{\sigma/\sqrt{n}}$ and is thus given by = NORMSDIST(z). For an upper test, the P-value is the area to the right of $\frac{\bar{x}-\mu_0}{\sigma/\sqrt{n}}$ and is given by = $1 -$ NORMSDIST(z). For a two-sided test, the formula for the P-value is = $2*(1 -$ NORMSDIST($|z|$)). The formula = ABS(z) returns the absolute value of z. The decision rule uses the logical = IF(statement, true, false), which returns the string designated as true if the statement is true or else it returns false.

Chapter 7

Inference for Distributions

7.1 Inference for the Mean of a Population

The workbook in Section 6.1 for the mean of a normal when σ is known can easily be modified to produce a level C confidence interval based on the Student t distribution,

$$\bar{x} \pm t^* \frac{s}{\sqrt{n}} \ .$$

As before n, $\bar{x}$ are the sample size and sample mean respectively, s is the sample standard deviation, and t^* is the critical t value such that the area between $-t^*$ and t^* under the curve of a t density with $n-1$ degrees of freedom equals C.

The Excel formula required is

$$= \texttt{TINV}(\alpha, \nu)$$

which returns the critical value for a level $C = 1 - \alpha$ confidence interval based on a t distribution with ν degrees of freedom.

The One-Sample t Confidence Interval

Example 7.1. (Example 7.1, page 495 in the text.) In fiscal 1996, the United States Agency for International Development provided 238,300 metric tons of corn soy blend (CSB) for development programs and emergency relief in countries throughout the world. CSB is a highly nutritious, low-cost fortified food that is partially precooked and can be incorporated into different food preparations by the recipients. As part of a study to evaluate appropriate vitamin C levels in this commodity, measurements were taken on samples of CSB produced in a factory. The following data are the amounts of vitamin C, measured in milligrams per 100 grams (mg/100 g) of blend, for a random sample of size 8 from a production run:

26   31   23   22   11   22   14   31

Find a 95% confidence interval for μ, the mean vitamin C content of the CSB produced during the run.

	A	B	C	D
1	T Interval for a Normal Mean – Values and Formulas			
2				
3		values	formulas	
4	User Input			
5	conf	0.95		Data
6	Summary Statistics			26
7	n	8	=COUNT(Data)	31
8	xbar	22.5	=AVERAGE(Data)	23
9	Calculations			22
10	s	7.19	=STDEV(Data)	11
11	SE	2.54	=s/SQRT(n)	22
12	df	7	=n-1	14
13	t	2.365	=TINV(1-conf, df)	31
14	ME	6.01	=t*SE	
15	Confidence Limits			
16	lower	16.49	=xbar-ME	
17	upper	28.51	=xbar+ME	

Figure 7.1: Confidence Interval for a Normal Mean

Solution. Create the workbook shown in Fig 7.1 using steps analogous to the production of Fig 6.1. In place of an assumed standard deviation σ, we calculate the sample standard deviation s using the Excel formula = STDEV(Data). The critical value t^* (denoted by t on the workbook) is obtained from the formula = TINV($1 - conf, df$), where $conf$ is the confidence level and $df = n - 1$ are the degrees of freedom. With this template in hand, it is only necessary to input and name the data. We have typed the eight data points in cells D6–D13 and **Named** them "Data" in cell D5. Column C has the formulas that you must enter in the respective cells in column B. The output provided by Excel gives a 95% confidence interval for μ as (16.49, 28.51) from cells B16:B17.

The One-Sample t Test

Again, the workbook based on the normal distribution in Section 6.2 is easily modified for use with Student's t distribution. For the significance test

$$H_0 : \mu = \mu_0$$

the test statistic is

$$t = \frac{\bar{x} - \mu_0}{s/\sqrt{n}}$$

which has Student's t distribution on $n - 1$ degrees of freedom. We remind the reader of the *unusual* Excel definition for the function that calculates the cumulative t, namely,

$$\text{TDIST}(x, \nu, 1) = P[t(\nu) > x]$$

where $t(\nu)$ is a t distribution on ν degrees of freedom. The argument x must be positive, and this accounts for the more complicated syntax in the decision rule in the corresponding template (Fig 7.2).

	A	B	C	D
1		T Test for a Normal Mean – Values and Formulas		
2				
3		values	formulas	
4	User Input			
5	null	40.0		Data
6	alpha	0.1		26
7				31
8	Summary Stats			23
9	n	8	=COUNT(Data)	22
10	xbar	22.5	=AVERAGE(Data)	11
11	Calculations			22
12	s	7.2	=STDEV(Data)	14
13	SE	2.542	=s/SQRT(n)	31
14	t	-6.883	=(xbar-null)/SE	
15	df	7	=n-1	
16	Lower Alternative			
17	lower_t		=-TINV(2*alpha, df)	
18	Decision		=IF(t<lower_t,"Reject H0", "Do Not Reject H0")	
19	Pvalue		=IF(t<0, TDIST(ABS(t),df,1), 1-TDIST(t,df,1))	
20	Upper Alternative			
21	upper_t		=TINV(2*alpha, df)	
22	Decision		=IF(t>upper_t,"Reject H0","Do Not Reject H0")	
23	Pvalue		=IF(t>0, TDIST(t,df,1), 1-TDIST(ABS(t),df,1))	
24	Two-Sided Alternative			
25	two_t	1.895	=TINV(alpha, df)	
26	Decision	Reject H0	=IF(ABS(t)>two_t,"Reject H0","Do Not Reject H0")	
27	Pvalue	0.00023	=TDIST(ABS(t), df, 2)	

Figure 7.2: One-Sample Student t Test

Example 7.2. (Example 7.2, page 496 in the text.) The specifications for the CSB described in Example 7.1 state that the mixture should contain 2 pounds of vitamin premix for every 2000 pounds of product. These specifications are designed to produce a mean (μ) vitamin C content in the final product of 40 mg/100 g. We test the null hypothesis that the mean vitamin C content is 40 mg/100 g.

$$H_0 : \mu = 40$$
$$H_a : \mu \neq 40$$

Solution. Fig 7.2 gives the formulas for carrying out a t test on a population mean. These are shown in column C but are to be entered in column B where Excel evaluates them. This problem involves a two-sided alternative, and column B shows only this output. The user inputs are the null value 40, $\alpha = 0.01$, and the data, entered in cells D6:D13. The calculated t value is -6.883, which is larger in absolute value than the 0.01 critical value for a two-sided test, namely 3.499. Therefore the conclusion is to reject H_0. The P-value is provided in cell B27. Its value of 0.00023 shows how extreme the evidence is against H_0.

Matched Pairs t Procedure

In order to reduce variability in a data set scientists sometimes use paired data matched on characteristics believed to affect the response. This is equivalent to a randomized block design. Such data are best analyzed if one-sample procedures are applied to differences between the pairs. There is a loss in degrees of freedom for error, but if the matching is effective, then this will be more than offset by the gain in reduced variance of the differences. The same method applies to before-after measurements on the same subjects.

Thus, we may apply the previous one-sample workbooks to the differences. Excel also provides a direct method for the matched pairs t test using the **Analysis ToolPak**. We will describe both approaches applied to the same data set.

Table 7.1: Pretest and Posttest Scores for French Immersion

Teacher	Pretest	Posttest	Teacher	Pretest	Posttest
1	32	34	11	30	36
2	31	31	12	20	26
3	29	35	13	24	27
4	10	16	14	24	24
5	30	33	15	31	32
6	33	36	16	30	31
7	22	24	17	15	15
8	25	28	18	32	34
9	32	26	19	23	26
10	20	26	20	23	26

Example 7.3. (Example 7.7, page 501 in the text.) The National Endowment for the Humanities sponsors a summer institute to improve the skills of high school teachers of foreign languages. One such institute hosted 20 French teachers for 4 weeks. At the beginning of the period, the teachers were given the Modern Languages Association's listening test of understanding of spoken French. After 4 weeks of immersion in French in and out of class, the listening test was given again. The pretest and posttest scores are provided in Table 7.1. Let μ denote the mean improvement that would be achieved if the entire population of French teachers attended a summer institute. We wish to test at the 0.01 level of significance

$$H_0 : \mu = 0$$
$$H_a : \mu > 0$$

	A	B	C	D	E	F
1			Paired T Test for a Normal Mean (Direct Approach)			
2						
3				Pretest	Posttest	Difference
4	User Input			32	34	2
5	null	0.0		31	31	0
6	alpha	0.01		29	35	6
7				10	16	6
8	Summary Stats			30	33	3
9	n	20	=COUNT(Difference)	33	36	3
10	xbar	2.5	=AVERAGE(Difference)	22	24	2
11	Calculations			25	28	3
12	s	2.9	=STDEV(Difference)	32	26	-6
13	SE	0.647	=s/SQRT(n)	20	26	6
14	t	3.865	=(xbar-null)/SE	30	36	6
15	df	19	=n-1	20	26	6
16	Upper Alternative			24	27	3
17	upper_t	2.539	=TINV(2*alpha, df)	24	24	0
18	Decision	Reject HO	=IF(t>upper_t,"Reject HO","Do Not Reject HO")	31	32	1
19	Pvalue	0.00052	=IF(t>0, TDIST(t,df,1), 1-TDIST(ABS(t),df,1))	30	31	1
20				15	15	0
21				32	34	2
22				23	26	3
23				23	26	3

Figure 7.3: Paired t Test—Direct Approach

To analyze these data, subtract the pretest scores from the posttest scores for each teacher. The 20 differences then form a single sample to which a one-sample test applies.

Matched Pairs t Test—Direct Approach

1. Adapt the template shown in Fig 7.2 by customizing it for this upper test (Fig 7.3).

2. Enter 0 for null and 0.01 for alpha, and record the data in a convenient place, say columns D, E, and F. Enter the label "Pretest" in cell D3 followed by the pretest scores in cells D4:D23. Enter the label "Posttest " in cell E3 followed by the posttest scores in cells E4:E23. Enter the label "Difference " in cell F3. Then in cell F4 enter the difference D4 – E4 in the **Formula Bar.** Select cell F4 and then move the mouse pointer over the fill handle in the lower right corner of cell F4. The pointer changes from an outline plus sign to a cross hair + when you are in position on the fill handle. Click the fill handle and drag it down so that the range F4:F23 is selected. Release the mouse button and Excel fills the contents of cells F4:F23 with the corresponding differences between posttest and pretest scores.

3. **Name** the difference range by selecting F3:F23, and from the Menu Bar choose **Insert – Name – Create** and then check the Top Row box in the **Create Names** dialog box. We can now refer to the data in F4:F23 by the

	A	B	C	D	E	F
1			T Test for a Normal Mean (Using the Analysis ToolPak)			
2						
3	Pretest	Posttest	Difference	t-Test: Paired Two Sample for Means		
4	32	34	2			
5	31	31	0		*Posttest*	*Pretest*
6	29	35	6	Mean	28.3	25.8
7	10	16	6	Variance	35.379	39.747
8	30	33	3	Observations	20	20
9	33	36	3	Pearson Correlation	0.890	
10	22	24	2	Hypothesized Mean Difference	0	
11	25	28	3	df	19	
12	32	26	-6	t Stat	3.865	
13	20	26	6	P(T<=t) one-tail	0.00052	
14	30	36	6	t Critical one-tail	2.539	
15	20	26	6	P(T<=t) two-tail	0.00104	
16	24	27	3	t Critical two-tail	2.861	
17	24	24	0			
18	31	32	1			
19	30	31	1			
20	15	15	0			
21	32	34	2			
22	23	26	3			

Figure 7.4: Paired *t* Test—Analysis ToolPak

name "Difference." (This is convenient but not necessary; we could equally use the cell reference F4:F23.)

4. Finally, where the name "Data" is used in formulas in the template of Fig 7.2, replace it with the name "Difference."

Excel Output

Fig 7.3 also gives the Excel output. The computed *t* value is 3.865, which exceeds the 1% critical value of 2.539. We therefore reject H_0. Additionally, we find that the *P*-value is 0.00052.

Matched Pairs *t* Test—Using the ToolPak

Excel also provides an **Analysis** tool for a matched pairs *t* test. However, *it cannot be used with summarized data*, while the direct approach can.

1. Open a new workbook and enter the unscented and scented values with their labels in cells A3:A23 and B3:B23, respectively, as in Fig 7.4.

2. In cell C3 enter the label "Difference." In cell C4 type the formula $= A4 - B4$. Then select C4 and fill to cells C4:C23.

3. Choose **Tools – Data Analysis** from the Menu Bar, and then check the box *t*-**Test:Paired Two Sample for Means**. Click OK.

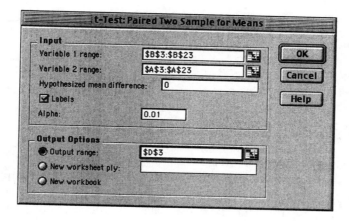

Figure 7.5: Paired t Test—Dialog Box

4. Complete the next dialog box shown in Fig 7.5. **Caution** is required in determining which scores are entered for **Variable 1 range** and which scores are entered for **Variable 2 range**. Excel takes Variable 1 - Variable 2 as the difference. Since we are taking posttest − pretest scores enter the posttest score range B3:B23 for Variable 1 and the pretest range A3:A23 for Variable 2.

Excel Output

The output appears in Fig 7.4 in the range D3:F16 beginning with cell D3, as specified in Fig 7.5. Individual sample means are given as well as the test statistic value t in cell E10. One-sided and two-sided critical values are provided in E14 and E16, as well as corresponding P-values in E13 and E15. Because $H_0 : \mu > 0$ and $t = 3.865$, we reject at level $\alpha = 0.01$. The P-value is 0.00052.

The entry $P(T <= t)$ one-tail in E13 is not $P(t(19) \leq 3.865)$, as the notation would suggest. Rather $P(T <= t)$ represents the tail area relative to t (so it is a lower tail if t is negative and an upper tail if t is positive). It is therefore not always a P-value.

Confidence Interval for Paired Data

The **Analysis ToolPak** does not provide a confidence interval directly, but it provides the information needed to carry out the calculations using the template in Section 7.1 applied to the differences. Again, adapt the workbook shown in Fig 7.1 for Example 7.1. In place of the "Data" range referenced in cells B7, B8, and B10 in Fig 7.1, point to the cells you have named "Difference," either with the direct approach or using the ToolPak. Excel will take the values from one workbook and use them in another. This is a reminder that data need not be on the same sheet or even the same workbook as your analysis.

> **Exercise.** For Example 7.3, show that a 90% confidence interval for the difference in posttest mean and pretest mean is (1.382, 3.618).

7.2   Comparing Two Means

Independent simple random samples of sizes n_1 and n_2 are obtained from populations with means μ_1 and μ_2, respectively. We are interested in comparing μ_1 and μ_2. The appropriate statistics and critical values required depend on the assumptions made. Suppose that the data are collected from normal populations with standard deviations σ_1 and σ_2. Denote by $\bar{x}_1, s_1^2, \bar{x}_2$, and s_2^2 the corresponding summary statistics (the sample means and sample variances).

Inference is usually based on a two-sample statistic of the form

$$\frac{(\bar{x}_1 - \bar{x}_2) - (\mu_1 - \mu_2)}{\text{SE}} \tag{7.1}$$

where SE represents the standard error of the numerator.

If the underlying populations are normal with known population standard deviations σ_1 and σ_2, then we use

$$z = \frac{(\bar{x}_1 - \bar{x}_2) - (\mu_1 - \mu_2)}{\sqrt{\frac{\sigma_1^2}{n_1} + \frac{\sigma_2^2}{n_2}}}$$

which has a standard normal distribution.

If, however, σ_1 and σ_2 are unknown, then the appropriate ratio is

$$t = \frac{(\bar{x}_1 - \bar{x}_2) - (\mu_1 - \mu_2)}{\sqrt{\frac{s_1^2}{n_1} + \frac{s_2^2}{n_2}}} \tag{7.2}$$

which is called a two-sample t statistic.

Actually, the distribution of the statistic in (7.2) depends on σ_1 and σ_2 and does not have an exact t distribution. Nonetheless, it is used with t critical values in inference in one of two ways, each involving a computed value for degrees of freedom ν associated with the denominator of (7.1) to provide an approximate t statistic. The two options follow.

1. Use a value for ν given by

$$\nu = \frac{\left(\frac{s_1^2}{n_1} + \frac{s_2^2}{n_2}\right)^2}{\frac{1}{n_1-1}\left(\frac{s_1^2}{n_1}\right)^2 + \frac{1}{n_2-1}\left(\frac{s_2^2}{n_2}\right)^2} \tag{7.3}$$

2. Use

$$\nu = \min\{n_1 - 1, n_2 - 1\} \tag{7.4}$$

Approximation (7.4) is conservative in the sense of providing a larger margin of error and is recommended in the text when doing calculations without the aid of software. Approximation (7.3) is considered to provide a quite accurate approximation to the actual distribution.

Excel, like most statistical software, uses the value for ν given in (7.3) and provides three tools in the **Analysis ToolPak** for analyzing independent samples from normal populations: z test, t test (unequal variances), and pooled t test (equal variances). These involve different sets of assumptions and consequently different forms for the SE. Each tool provides a similar dialog box for user input, parameters, and the range for the actual raw data.

However, sometimes only summary statistics are available and the **Analysis ToolPak** cannot be used; instead, direct calculations are required.

Two-Sample z Statistic

For normal populations with known standard deviations, the "standard score"

$$z = \frac{(\bar{x}_1 - \bar{x}_2) - (\mu_1 - \mu_2)}{\sqrt{\frac{\sigma_1^2}{n_1} + \frac{\sigma_2^2}{n_2}}}$$

has a standard normal distribution. This ratio is also used for "large sample" procedures with σ_1 and σ_2 unknown and replaced by the sample standard deviations s_1 and s_2. Normality of the underlying population is not mandatory for large sample tests, and the corresponding confidence intervals and significance tests have approximately the specified values.

Two-Sample z Confidence Interval

Example 7.4. (Exercise 7.70(b), page 548 in the text.) Does cocaine use by pregnant women cause their babies to have low birth weight? To study this question, birth weights of babies of women who tested positive for cocaine/crack during a drug-screening test were compared with the birth weights for women who either tested negative or were not tested, a group we call "other." Here are the summary statistics. The birth weights are measured in grams.

Group	n	$\bar{x}$	s
Positive test	134	2733	599
Others	5974	3118	672

Give a 95% confidence interval for the mean difference in birth weights.

	A	B	C	D
1	Two-Sample Z Confidence Interval			
2				
3		Summary Statistics and User Input		
4	Group	n	xbar	s
5	Positive	134	2733	599
6	Other	5974	3118	672
7				
8	conf	0.95		
9	SE	52.471	=SQRT(D5^2/B5+D6^2/B6)	
10	z	1.960	=NORMSINV(0.5+conf/2)	
11	ME	102.841	=z*SE	
12	Confidence Limits			
13	lower	-487.84	=(C5-C6)-ME	
14	upper	-282.16	=(C5-C6)+ME	

Figure 7.6: Large Two-Sample z Confidence Interval

Solution. We will use a procedure based on a large sample z, treating the observed sample standard deviations as the corresponding population values. Fig 7.6 provides both formulas (column C) to derive the required confidence interval and the values obtained (column B). The user inputs are $\bar{x}_1$, $\bar{x}_2$, σ_1, σ_2, n_1, n_2, and the confidence level C. We have **Named** the ranges to refer to "conf," "SE," "z," and "ME." We can read off from cells B13:B14 that a 95% confidence interval is $(-487.84, -282.16)$.

Two-Sample z Test

> **Example 7.5.** (Continuing Exercise 7.70, page 548 in the text.) Test the hypothesis that babies of women who use cocaine while pregnant have a lower birth weight on average than babies of women in the "other" group.

Solution. The hypothesis to be tested is

$$H_0 : \mu_1 = \mu_2$$
$$H_a : \mu_1 < \mu_2$$

where μ_1 and μ_2 are the hypothesized mean birth weights for babies born to women who test positive and the "others," respectively. At the level $\alpha = 0.05$, the decision rule is to reject H_0 if the computed z value

$$z = \frac{\bar{x}_1 - \bar{x}_2}{\sqrt{\frac{\sigma_1^2}{n_1} + \frac{\sigma_2^2}{n_2}}}$$

satisfies $z < 1.645$. The P-value will be given by

$$P\text{-value} = \Phi(z)$$

where $\Phi(z)$ is the cumulative $N(0,1)$ distribution function. As in the previous example, in view of the large sample sizes, we can use s_1 and s_2, the sample standard deviations, in place of σ_1 and σ_2. These calculations can readily be made by a hand calculator.

	A	B	C	D	F
1		Large Two-Sample Z Test			
2					
3		Summary Statistics and User Input			
4	Group	n	xbar	s	
5	Positive	134	2733	599	
6	Other	5974	3118	672	
7					
8	SE	52.471	=SQRT(D5^2/B5+D6^2/B6)		
9	z	-7.337	=((C5-C6)-null)/SE		
10	alpha	0.05			
11	null	0			
12	Lower Test				
13	lower_z	-1.645	=NORMSINV(alpha)		
14	Decision	Reject HO	=IF(z<lower_z,"Reject HO","Do Not Reject HO")		
15	Pvalue	1.097E-13	=NORMSDIST(z)		
16	Upper Test				
17	upper_z		=-NORMSINV(alpha)		
18	Decision		=IF(z>upper_z,"Reject HO","Do Not Reject HO")		
19	Pvalue		=1-NORMSDIST(z)		
20	Two-Sided Test				
21	two_z		=ABS(NORMSINV(alpha/2))		
22	Decision		=IF(ABS(z)>two_z,"Reject HO","Do Not Reject HO")		
23	Pvalue		=2*(1-NORMSDIST(ABS(z)))		

Figure 7.7: Large Two-Sample z Test

While the full power of Excel comes from dealing with large data sets, even this simple example can illustrate the use of a spreadsheet to evaluate formulas within equations. Fig 7.7 is an Excel workbook containing a template for this type of problem showing the Excel formulas required. The calculations are carried out in column B and the corresponding formulas are displayed in the adjacent cells in column C. The problem at hand is a lower test, but formulas are additionally provided for both upper and two-sided tests, only one of which should be used at any time. For purposes of clarity, we have used **Named Ranges** rather than cell references to refer to $\alpha = $ alpha in cell B10, to z in B9, to the standard error

$$ SE = \sqrt{\frac{s_1^2}{n_1} + \frac{s_2^2}{n_2}} $$

in B8, and to the critical values lower_z, upper_z, and two_z for the three types of alternative hypotheses. Remember to name your ranges when copying and adapting this template to your own workbook.

The Excel output appears in Fig 7.7 in cells B8:B15 and the lower 5% critical value of -1.645 is in B13. The conclusion "Reject H_0" is printed by Excel in B14, and the P-value, 1.097×10^{-13}, is also evaluated in B15.

Two-Sample z Inference Using the ToolPak

While the aforementioned templates have been designed for summarized data, they can easily be modified for use with raw data, for instance, by entering = AVERAGE(Range) in place of the *value* of the sample mean, where Range is the cell range for the sample. Since population standard deviations are seldom known but instead are replaced by their sample estimates in most applications, if you are dealing with large raw data sets whose standard deviations have not yet been calculated, it is simplest to then let Excel do the calculation. So enter = STDEV(Range) where you would enter the value for the standard deviation and then proceed as before. Excel computes the standard deviation from the data and inserts it where required in the formulas.

There is a two-sample z test option included in the **Analysis ToolPak**. But for large sample sizes, the results of two-sample t and z options are virtually the same and the sequence of steps in the two tools are identical. Therefore, as we will discuss the two-sample t ToolPak in detail in the next section, we will not illustrate its z counterpart.

Two-Sample t Procedures

Using Summarized Data

Suppose only $\bar{x}_1, s_1^2, \bar{x}_2$, and s_2^2, not the raw data, are available. We will use Excel to calculate (7.2) and use it for a test of significance and a confidence interval.

> **Example 7.6.** (Example 7.14, page 530 in the text.) An educator believes that new directed reading activities in the classroom will help elementary school pupils improve some aspects of their reading ability. She arranges for a third-grade class of 21 students to take part in these activities for an eight-week period. A control classroom of 23 third graders follows the same curriculum with the activities. At the end of the eight weeks, all students are given a Degree of Reading Power (DRP) test, which measures the aspects of reading ability that the treatment is designed to improve. The summary statistics are
>
> | Group | n | $\bar{x}$ | s |
> |---|---|---|---|
> | Treatment | 21 | 51.48 | 11.01 |
> | Control | 23 | 41.52 | 17.15 |
>
> Carry out the significance test at the level $\alpha = 0.05$.

Solution. Because we hope to show that the treatment (group 1) is better than the control (group 2), the hypotheses are

$$H_0 : \mu_1 = \mu_2$$
$$H_a : \mu_1 > \mu_2$$

	A	B	C	D	F	G
1	Two–Sample T Test Summarized Data					
2						
3		Summary Statistics and User Input				
4	Group	n	xbar	s		
5	Treatment	21	51.48	11.01		
6	Control	23	41.52	17.15		
7						
8	SE	4.308	=SQRT(D5^2/B5+D6^2/B6)			
9	t	2.312	=((C5-C6)-null)/SE			
10	alpha	0.05				
11	null	0				
12	mindf	20	=MIN(B5-1,B6-1)			
13	numerator	344.485	=POWER((D5^2/B5+D6^2/B6),2)			
14	denominator	9.099	=(D5^4/B5^2)/(B5-1)+(D6^4/B6^2)/(B6-1)			
15	df	38	=1+INT(B13/B14)			
16	Lower Test					
17	lower_t		=-TINV(2*alpha, df)			
18	Decision		=IF(t<lower_t,"Reject H0", "Do Not Reject H0")			
19	Pvalue		=IF(t<0, TDIST(ABS(t),df,1), 1-TDIST(t,df,1))			
20	Upper Test					
21	upper_t	1.686	=TINV(2*alpha, df)			
22	Decision	Reject H0	=IF(t>upper_t,"Reject H0","Do Not Reject H0")			
23	Pvalue	0.0131	=IF(t>0, TDIST(ABS(t),df,1), 1-TDIST(ABS(t),df,1))			
24	Two–Sided Test					
25	two_t		=TINV(alpha, df)			
26	Decision	Reject H0	=IF(ABS(t)>two_t,"Reject H0","Do Not Reject H0")			
27	Pvalue		=TDIST(ABS(t), df, 2)			

Figure 7.8: Two-Sample t Test—Summarized Data

At the level $\alpha = 0.05$, reject H_0 if the computed t value

$$t = \frac{\bar{x}_1 - \bar{x}_2}{\sqrt{\frac{s_1^2}{n_1} + \frac{s_2^2}{n_2}}}$$

satisfies $|t| > t^*$, where t^* is the upper $\alpha/2 = 0.025$ critical value of a t distribution.

As in Example 7.5, we have provided an Excel workbook, Fig 7.8, containing a template for this type of problem, where the calculations are carried out in column B while the actual formulas in the cells behind column B are given in the adjacent cells in column C. We are dealing with an upper-level test, but formulas are also provided for lower and two-sided alternatives.

Cell B8 gives the standard error (the denominator of (7.2)). Cell B12 gives the conservative degrees of freedom, 20, for the value of ν given by (7.4), while cells B13:B15 calculate the degrees of freedom, 38, corresponding to the value of ν in (7.3). The calculated t statistic in cell D9 is 2.312. Using the more accurate degrees of freedom, 38, the corresponding critical t value of 1.686 at level .05 is presented in cell B21. We therefore reject H_0 at level $\alpha = 0.05$. Observe that the calculated P-value is given in cell B23 in Fig 7.8 as 0.0131.

Example 7.7. Find a 95% confidence interval for the mean improvement in the entire population of third graders in Example 7.6.

	A	B	C	D	E
1		Two–Sample T CI Summarized Data			
2					
3		Summary Statistics and User Input			
4	Group	n	xbar	s	
5	Treatment	21	51.48	11.01	
6	Control	23	41.52	17.15	
7					
8	conf	0.95	0.95		
9	mindf	20	=MIN(C5-1,C6-1)		
10	numerator	344.30	=POWER((E5^2/C5+E6^2/C6),2)		
11	denominator	9.10	=(E5^4/C5^2)/(C5-1)+(E6^4/C6^2)/(C6-1)		
12	df	38	=1+INT(C10/C11)		
13	Calculations				
14	SE	4.308	=SQRT(E5^2/C5+E6^2/C6)		
15	crit_t	2.024	=TINV(1-conf, df)		
16	ME	8.720	=crit_t*SE		
17	lower	1.23	=(D5-D6)-ME		
18	upper	18.67	=(D5-D6)+ME		

Figure 7.9: Two-Sample t Confidence Interval

Solution. We modify the workbook for the z procedure given in Fig 7.6 to derive a two-sample t confidence interval. The confidence interval is given by

$$\bar{x}_1 - \bar{x}_2 \pm t^* \sqrt{\frac{s_1^2}{n_1} + \frac{s_2^2}{n_2}}$$

where t^* is a critical t-value on an appropriate number of degrees of freedom. The formulas required and values are shown in Fig 7.9. We find that a 95% confidence interval based on 38 degrees of freedom is (1.23, 18.67).

Using the Analysis ToolPak

As mentioned, Excel provides three tools in the **Analysis ToolPak** for comparing means from two populations based on independent samples. These are the two-sample t test, pooled two-sample t test, and two-sample z test. These tools provide dialog boxes in which the user locates the data and decides on the type of analysis desired. For large sample sizes, the results of two-sample t and z options are virtually the same and the sequence of steps in the two tools are identical. Therefore, as we will be discussing the two-sample t ToolPak in detail, we will not illustrate its z counterpart. Besides, there is a bug in the two-sample z ToolPak that outputs an incorrect two-sided P-value; moreover, there is no built-in confidence interval procedure. These considerations limit the usefulness of the two-sample z ToolPak, and we do not recommend using it.

We illustrate use of the ToolPak with Examples 7.6 and 7.7 not only for comparison purposes but also because the full data set appears as Table 7.2 in the text and may be readily referenced.

Example 7.8. Redo Examples 7.6 and 7.7 with the raw data using the **Analysis ToolPak**.

Solution

1. Open a new workbook and enter the data from Table 7.3 on page 530 of the text. Insert the treatment group in cells A4:A24 and the control group in Cells B4:B26. Enter the label "Treatment" in A3 and the label "Control" in B3. (Refer to Fig 7.12, which also shows the output.)

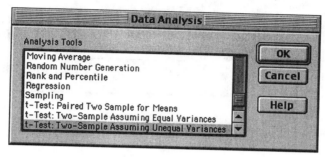

Figure 7.10: Two-Sample *t* Test Analysis Tools

2. From the Menu Bar choose **Tools – Data Analysis** and select *t*-**Test: Two-Sample Assuming Unequal Variances** from the list of selections (Fig 7.10). Click OK (equivalently, double-click your selection).

3. A dialog box (Fig 7.11) appears. Complete as shown. **Variable 1 range** refers to the cell addresses of the sample you have designated by subscript 1, in this case the treatment group. Type A3:A24 in its text area (with the flashing vertical I-beam). Alternatively, you can point to the data by clicking on cell A3 and dragging to the end of the treatment data, cell A24. The values $A$3:$A$24 will appear in the text area of the dialog box. Similarly, enter the range B3:B26 for the control group in the **Variable 2 range**.

4. The **Hypothesized mean difference** refers to the null value, which is 0 here. Check the **Labels** box, because your ranges included the labels for the two groups. The level of significance **Alpha** is the default 0.05. We will be placing the output in the same workbook as the data, so check the radio button **Output range** and type C3 in the text area. Finally, click OK. The output will appear in a block of cells whose upper left corner is cell C3 in Fig 7.12.

Excel Output

The output appears in the range C3:E15 in Fig 7.12. The range C18:E23 is not part of the output but is the result of additional formulas we have entered to give confidence intervals (discussed next). From cells D6:E6, we see that the sample

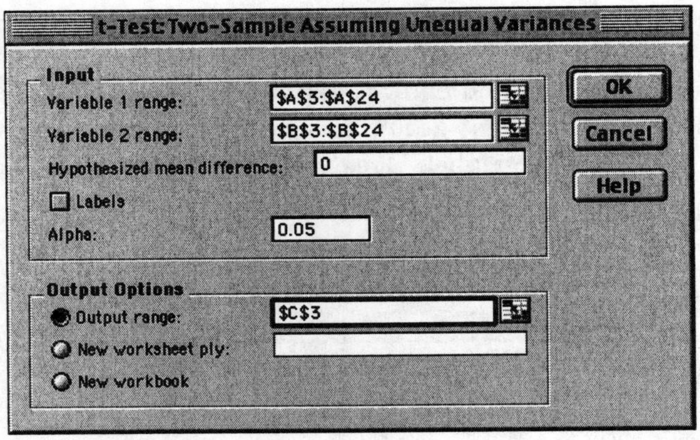

Figure 7.11: Two-Sample t Test Dialog Box

means for treatment and control groups are 51.476 and 41.522, while the sample variances are 121.162 and 294.079, respectively. The degrees of freedom are 38 (Excel rounds up to the nearest integer). The computed t statistic in cell D11 is 2.311, while the one-sided critical t^* value on 38 degrees of freedom at the 5% level in cell D13 is 1.686. Since the computed t exceeds t^*, we reject the null hypothesis and conclude that there is strong evidence that the directed reading activities help elementary school pupils improve some aspects of their reading ability.

Excel provides P-values in cells D12 and D14. Here the P-value is 0.0132. As before, the entry "$P(T <= t)$ one-tail" in C12 needs some explanation. It is meant to be a one-tailed P-value, which depends on the calculated t Stat, the computed t statistic. If t Stat < 0, then $P(T <= t)$ one-tail is in fact the lower tail corresponding to the area to the left of t Stat under a t density curve. But if t is positive, then $P(T <= t)$ one-tail is the area to the right of t Stat. It is therefore not always the P-value. For instance, if the test to be carried out were

$$H_0 : \mu_1 = \mu_2$$
$$H_a : \mu_1 < \mu_2$$

then the P-value would be $1 - 0.0132 = 0.9868$ rather than 0.0121. $P(T <= t)$ two-tail is the correct P-value for a two-tailed test.

Confidence Intervals

The **ToolPak** does not print a confidence interval directly, but the output provides enough information to carry out the calculations. Details are given in cells C18:E23 of Fig 7.12. The cells in column E of this block show the formulas that are the entries behind the cells in column D and whose values are evaluated and printed in

	A	B	C	D	E	F
1			**Two-Sample t Test**			
2						
3	Treatment	Control	t-Test: Two-Sample Assuming Unequal Variances			
4	24	10				
5	33	17		*Treatment*	*Control*	
6	43	19	Mean	51.476	41.522	
7	43	20	Variance	121.162	294.079	
8	43	26	Observations	21	23	
9	44	28	Hypothesized Mean Difference	0		
10	46	33	df	38		
11	49	37	t Stat	2.311		
12	49	37	P(T<=t) one-tail	0.013		
13	52	41	t Critical one-tail	1.686		
14	53	42	P(T<=t) two-tail	0.026		
15	54	42	t Critical two-tail	2.024		
16	56	42				
17	57	43				
18	57	46	Mean Difference	9.954	=D6-E6	
19	58	48	SE	4.308	=SQRT(D7/D8+E7/E8)	
20	59	53	t	2.024	=TINV(0.05, D10)	
21	61	54	ME	8.72	=D20*D19	
22	62	55	lower	1.23	=D18-D21	
23	67	55	upper	18.67	=D18+D21	
24	71	60				
25		62				
26		85				

Figure 7.12: Two-Sample t Test ToolPak Output

the workbook by Excel. These formulas are the Excel equivalents of the formula

$$\bar{x}_1 - \bar{x}_2 \pm t^* \sqrt{\frac{s_1^2}{n_1} + \frac{s_2^2}{n_2}}$$

The information needed—the sample means, sample variances, sample sizes, and the critical t^* values—is part of the **ToolPak** output and is referenced in cells C18:E23. As before (Fig 7.9, from Example 7.7), the 95% confidence interval is (1.23, 18.67).

The Pooled Two-Sample t Procedures

When the two populations are believed to be normal with the same variance, it is more common to use a pooled two-sample t based on an exact t distribution. The procedures are based on the statistic

$$t = \frac{(\bar{x}_1 - \bar{x}_2) - (\mu_1 - \mu_2)}{s_p \sqrt{\frac{1}{n_1} + \frac{1}{n_2}}} \tag{7.5}$$

where $s_p^2 = \frac{(n_1-1)s_1^2 + (n_2-1)s_2^2}{n_1+n_2-2}$ is called the pooled sample variance. This statistic is known to have an exact t distribution on $\nu = n_1 + n_2 - 2$ degrees of freedom. The previous two-sample t analyses carry over with the obvious modifications for the degrees of freedom and use of the denominator in (7.5) in place of the denominator in (7.2).

Example 7.9. (Examples 7.19, 7.20, and 7.21, pages 539–545 in the text.) Does increasing the amount of calcium in our diet reduce blood pressure? A randomized comparative experiment gave one group of 10 black men a calcium supplement for 12 weeks. The control group of 11 black men received a placebo that appeared identical.

(a) Test the hypothesis that calcium lowers blood pressure more than a placebo by testing

$$H_0 : \mu_1 = \mu_2$$
$$H_a : \mu_1 > \mu_2$$

at level $\alpha = 0.05$.

(b) Estimate the effect of calcium supplementation by computing a 90% confidence interval for the difference in population means $\mu_1 - \mu_2$.

Solution

1. Enter the data and labels in A3:A13 (Calcium) and B3:B14 (Placebo) of a workbook (Fig 7.13).

2. From the Menu Bar, choose **Tools – Data Analysis** and select *t*-**Test: Two-Sample Assuming Equal Variances** from the list of selections. (Refer to the dialog box in Fig 7.10.) Click OK.

3. Complete the next dialog box, which is similar to Fig 7.11, exactly as you did for the unequal variances case.

Excel Output

The output appears in the range C1:E14 in Fig 7.13, and we see that

$$\bar{x}_1 = 5.000 \qquad s_1^2 = 76.444$$
$$\bar{x}_2 = -0.273 \qquad s_2^2 = 34.818$$
$$s_p^2 = 54.536$$

The computed pooled t statistic on 19 degrees of freedom is $t = 1.634$ (cell D12) while the $\alpha = 0.05$ upper critical value is $t^* = 1.729$ (cell D14). We conclude that although the calcium supplement appears to lower blood pressure, the difference between the calcium group and the placebo group is not significant at the 5% level. The P-value 0.0593 is shown in cell D13.

	A	B	C	D	E	F	G
1			**Pooled Two-Sample t Test**				
2							
3	Calcium	Placebo	t-Test : Two-Sample Assuming Equal Variances				
4	7	-1					
5	-4	12			*Calcium*	*Placebo*	
6	18	-1	Mean	5.000	-0.273		
7	17	-3	Variance	76.444	34.818		
8	-3	3	Observations	10	11		
9	-5	-5	Pooled Variance	54.536			
10	1	5	Hypothesized Mean Difference	0			
11	10	2	df	19			
12	11	-11	t Stat	1.634			
13	-2	-1	P(T<=t) one-tail	0.0593			
14		-3	t Critical one-tail	1.729			
15			P(T<=t) two-tail	0.119			
16			t Critical two-tail	2.093			
17							
18			Mean Difference	5.273	=D6-E6		
19			SE	3.227	=SQRT(D9)*SQRT((1/10 + 1/11))		
20			t	1.729	=TINV(0.1 , D11)		
21			ME	5.579	=D20*D19		
22			lower	-0.307	=D18-D21		
23			upper	10.852	=D18+D21		

Figure 7.13: Two-Sample t ToolPak Output

Confidence Intervals

In order to supplement the **Analysis ToolPak** output, which does not provide a confidence interval, we have also included in C18:D23 of Fig 7.13 formulas and output for the confidence intervals. These rely on the summary statistics produced by the ToolPak and are the Excel equivalents of the formula

$$\bar{x}_1 - \bar{x}_2 \pm t^* s_p \sqrt{\frac{1}{n_1} + \frac{1}{n_2}}$$

From D22:D23 we read a 90% confidence interval $(-0.307, 10.852)$.

7.3 Optional Topics in Comparing Distributions

Inference for Population Spread

Suppose that s_1^2 and s_2^2 are the sample variances of independent simple random samples of sizes n_1 and n_2 taken from normal populations $N(\mu_1, \sigma_1)$ and $N(\mu_2, \sigma_2)$, respectively. Then the ratio

$$F = \frac{s_1^2/\sigma_1^2}{s_2^2/\sigma_2^2}$$

has a known sampling distribution that does not depend on $\{\mu_1, \mu_2, \sigma_1, \sigma_2\}$, but only on the sample sizes. It has an F distribution on $n_1 - 1$ and $n_2 - 1$ degrees of freedom for the numerator and the denominator, respectively. The ratio on the right side of the equation is only one manifestation of the F distribution, which is also used in analysis of variance and regression.

The F Test for Equality of Spread

In this section the context is comparison of σ_1 and σ_2. It turns out for mathematical reasons that the appropriate parameter for testing the null hypothesis

$$H_0 : \sigma_1 = \sigma_2$$

is the ratio $\frac{\sigma_1}{\sigma_2}$ (equivalently $\frac{\sigma_1^2}{\sigma_2^2}$) rather than the difference, which we used for comparing means.

Example 7.10.   (Example 7.19, page 539 in the text.) Does increasing the amount of calcium in our diet reduce blood pressure? Examination of a large sample of people revealed a relationship between calcium intake and blood pressure. A randomized comparative experiment gave one group of 10 black men a calcium supplement for 12 weeks. The control group of 11 black men received a placebo that appeared identical. Table 7.2 gives the seated systolic (heart contracted) blood pressure for all subjects at the beginning and end of the 12-week period, in millimeters of mercury. The analysis employed a pooled two-

Table 7.2: Seated Systolic Blood Pressure

Calcium group			Placebo group		
Begin	End	Decrease	Begin	End	Decrease
107	100	7	12	124	−1
110	114	−4	109	97	12
123	105	18	112	113	−1
129	112	17	102	105	−3
112	115	−3	98	95	3
111	116	−5	114	119	−5
107	106	1	119	114	5
112	102	10	112	114	2
136	125	11	110	121	−11
102	104	−2	117	118	−1
			130	133	−3

sample t test, which required assumption of equal population variances. Using level $\alpha = 0.05$ test

$$H_0 : \sigma_1 = \sigma_2$$
$$H_a : \sigma_1 \neq \sigma_2$$

	A	B	C	D	E	F
1			F Test	for Equality	of Variances	
2	Differences in Systolic BP					
3	Calcium	Placebo				
4	7	-1		F-Test Two-Sample for Variances		
5	-4	12				
6	18	-1				
7	17	-3			Calcium	Placebo
8	-3	3		Mean	5.00	-0.27
9	-5	-5		Variance	76.444	34.818
10	1	5		Observations	10	11
11	10	2		df	9	10
12	11	-11		F	2.1955	
13	-2	-1		P(F<=f) one-tail	0.1182	
14		-3		F Critical one-tail	3.7790	

Figure 7.14: F Test Data and Output

Solution. We will use the F test in the **Analysis ToolPak**.

1. Enter the data from Table 7.2 in columns A and B of a workbook (Fig 7.14).

2. From the Menu Bar choose **Data Analysis – F-Test Two-Sample for Variances** and complete the dialog box of Fig 7.15. Notice that we have inserted not the specified level of significance $\alpha = 0.05$ in this box but rather half the value, 0.025, to reflect the fact that our test is two-sided while the cells E12:E13 give the P-value and the critical value for a one-sided upper-tailed test.

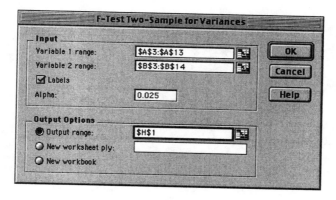

Figure 7.15: F Test Dialog Box

Caveat

This tool requires that the larger of the two sample variances be in the numerator, so repeat this procedure by reversing the variables if the output shows that the variance of the data in **Variable 1 range** in Fig 7.15 is less than that in **Variable 2 range** (which is not the case here).

Excel Output

The output appears in cells D4:F13, as shown in Fig 7.14. The computed value of F under H_0

$$F = \frac{s_1^2}{s_2^2} = 2.1955$$

appears in E11. (Note that $s_1^2 > s_2^2$ as required.) The critical F value is 3.779 and therefore the data are not significant at the 5% level.

We can obtain from cell E12 a one-sided P-value, which we need to double and find that the P-value $= 0.236$.

The F Distribution Function

This a good place to record the syntax for the F distribution. Suppose that F is a random variable having an F distribution with degrees of freedom ν_1 for the numerator and ν_2 for the denominator. Then for any $x > 0$,

$$\texttt{FDIST}(x, \nu_1, \nu_2) = P(F > x)$$

while for any $0 < p < 1$, the upper p critical value is obtained from the inverse

$$P\left(F > \texttt{FINV}(p, \nu_1, \nu_2)\right) = p$$

Chapter 8

Inference for Proportions

In this chapter we discuss data representing the counts or proportions of outcomes occurring in different categories in a population.

8.1   Inference for a Single Proportion

To estimate the proportion p of some characteristic in a population, it is common to take an SRS of size n and count $X =$ the number in the sample possessing the characteristic. For large n, the distribution of X is approximately binomial $B(n, p)$, and by the central limit theorem the sample proportion

$$\hat{p} \text{ is approximately } N\left(p, \sqrt{\frac{p(1-p)}{n}}\right)$$

Inference is then based on the procedures for estimating a normal mean discussed in Chapter 6.

Confidence Intervals

The standard error of $\hat{p}$ is

$$SE_{\hat{p}} = \sqrt{\frac{\hat{p}(1-\hat{p})}{n}}$$

where we have replaced p with $\hat{p}$ in the expression for the standard deviation of $\hat{p}$. Therefore a large-sample level C confidence interval for p is given as

$$\hat{p} \pm z^* SE_{\hat{p}}$$

where z^* is the upper $(1-C)/2$ standard normal critical value.

Example 8.1. (Example 8.1, page 574 in the text.) Alcohol abuse has been described by college presidents as the number one problem on campus, and it is an important cause of death in young adults. How common is it? A survey of 17,096 students in four-year colleges collected information on drinking behavior and alcohol-related problems. The researchers defined *frequent binge drinking* as having five or more drinks in a row, three or more times in the past two weeks. According to this definition, 3,314 students were classified as frequent binge drinkers. The proportion of drinkers is

$$\hat{p} = \frac{3314}{17096} = 0.194$$

Find a 95% confidence interval for the proportion of binge drinkers among all four-year college students.

	A	B	C
1		Confidence Interval for a Proportion	
2			
3		Values	Formulas
4	User Input		
5	conf	0.95	
6	Summary Statistics		
7	n	17096	
8	X	3314	
9	Calculations		
10	p_hat	0.1938	=X/n
11	SE	0.0030	=SQRT(p_hat*(1-p_hat)/n)
12	z	1.96	=NORMSINV(0.5+conf/2)
13	ME	0.006	=z*SE
14	Confidence Limits		
15	lower	0.188	=p_hat-ME
16	upper	0.200	=p_hat+ME

Figure 8.1: Confidence Interval for a Population Proportion

Solution.   Fig 8.1 shows the Excel formulas required for the calculation (in column C), together with the corresponding values obtained when these formulas are entered in column B on your workbook. The formulas parallel those for the confidence interval for a normal mean. From cells B15:B16 we find that a 95% confidence interval is (0.188, 0.200).

Note: We remind you that the formulas in Fig 8.1 require **Named Ranges** to refer to the variables by their names, or else the cell references must be used.

Significance Tests

For testing the null hypothesis

$$H_0 : p = p_0$$

we use the test statistic

$$z = \frac{\hat{p} - p_0}{\sqrt{\frac{p_0(1-p_0)}{n}}}$$

for a large-sample procedure. For example, if the alternative is two-sided

$$H_a : p \neq p_0$$

then we reject H_0 at level α if

$$|z| > z^*$$

where z^* is the upper $\frac{\alpha}{2}$ standard normal critical value. Furthermore the P-value is $2P(Z > |z|)$, where Z is $N(0,1)$.

	A	B	C
1		**Significance Test for a Proportion**	
2			
3	User Input		
4	p0	0.5	
5	alpha	0.01	
6	alternate	two-sided	
7	Summary Statistics		
8	n	4040	
9	X	2048	
10	Calculations		
11	p_hat	0.5069	=X/n
12	SE	0.0079	=SQRT(p0*(1-p0)/n)
13	z	0.881	=(p_hat-p0)/SE
14	Lower Test		
15	lower_z		=NORMSINV(alpha)
16	Decision		=IF(z<lower_z,"Reject H0","Do Not Reject H0")
17	Pvalue		=NORMSDIST(z)
18	Upper Test		
19	upper_z		=-NORMSINV(alpha)
20	Decision		=IF(z>upper_z,"Reject H0","Do Not Reject H0")
21	Pvalue		=1-NORMSDIST(z)
22	Two-Sided Test		
23	two_z	2.576	=ABS(NORMSINV(alpha/2))
24	Decision	Do Not Reject H0	=IF(ABS(z)>two_z,"Reject H0","Do Not Reject H0")
25	Pvalue	0.378	=2*(1-NORMSDIST(ABS(z)))

Figure 8.2: Significance Test for a Population Proportion

Example 8.2. (Example 8.2, page 576 in the text.) The French naturalist Count Buffon once tossed a coin 4040 times and obtained 2048 heads. To assess whether the data provide evidence that the coin was not balanced, we test

$$H_0 : p = 0.5$$
$$H_a : p \neq 0.5$$

where p is the probability that Buffon's coin lands heads.

Solution. Fig 8.2 gives the required formulas in column C for lower, upper, and two-sided tests. Column B contains the data and calculations. Cell B11 gives $\hat{p} = 0.5069$, cell B12 gives the standard error SE = 0.0079, and cell B13 gives the calculated value of $z = 0.881$. Our problem is two-sided, so only the values in rows

22–25 of the template are relevant. The P-value is 0.378, and we therefore do not reject H_0.

These calculations are straightforward, and Fig 8.2 merely provides a systematic way of carrying them out. Nothing so elaborate is really needed for calculations that could be done with a hand calculator.

> **Exercise.** (Adapted from Example 8.4, page 577 in the text.) Simulate 4040 tosses of a fair coin. Repeat the simulation 100 times and plot the lower and upper end points of the 100 confidence intervals on a graph with the constant line $p = 0.5$. How many intervals contain 0.5?

8.2   Comparing Two Proportions

Large-sample inference procedures for comparing the proportions p_1 and p_2 in two populations based on independent SRS of sizes n_1 and n_2, respectively, are also based on the normal approximation. The natural estimate $D = \hat{p}_1 - \hat{p}_2$ of the difference in proportions $p_1 - p_2$ is approximately normal with mean $p_1 - p_2$ and standard deviation

$$\sigma = \sqrt{\frac{p_1(1-p_1)}{n_1} + \frac{p_2(1-p_2)}{n_2}}$$

Confidence Intervals

We must replace the unknown parameters p_1 and p_2 by their estimates $\hat{p}_1$ and $\hat{p}_2$ to obtain an estimated standard error

$$\mathrm{SE}_D = \sqrt{\frac{\hat{p}_1(1-\hat{p}_1)}{n_1} + \frac{\hat{p}_2(1-\hat{p}_2)}{n_2}}$$

and an approximate level C confidence interval

$$\hat{p}_1 - \hat{p}_2 \pm z^* \mathrm{SE}_D$$

where z^* is the upper $(1-C)/2$ standard normal critical value.

> **Example 8.3.**   (Example 8.8, page 589 in the text.) This example continues the binge drinking study of Example 8.1 by examining whether there are differences according to gender. The table at the bottom of page 589 in the text provides the sample proportions of frequent binge drinkers separately for men and for women in the column labeled " $\hat{p}$." The last line gives the totals which were used in Example 8.1. Find a 95% confidence interval for the difference between the proportions of men and women who are frequent binge drinkers.

Solution. The data is shown in the Excel worksheet in Fig 8.3. Results of the calculations are given in column B with the corresponding formulas in the adjacent column C. Based on this data set we are 95% confident that the difference in proportions between men and women is in the range (0.045, 0.069).

	A	B	C	D	E
1		Confidence Interval for a Difference in Proportions			
2					
3		Summary Statistics and User Input			
4	Group	n	X	p_hat=X/n	
5	Men	7180	1630	0.2270	
6	Women	9916	1684	0.1698	
7					
8	conf	0.95			
9	SE	0.00622	=SQRT((D5*(1-D5)/B5)+(D6*(1-D6)/B6))		
10	z	1.960	=NORMSINV(0.5+conf/2)		
11	ME	0.012	=z*SE		
12	Confidence Limits				
13	lower	0.045	=(D5-D6)-ME		
14	upper	0.069	=(D5-D6)+ME		

Figure 8.3: Confidence Interval—Difference in Proportions

Significance Tests

The null hypothesis

$$H_0 : p_1 = p_2$$

is tested using the statistic

$$z = \frac{\hat{p}_1 - \hat{p}_2}{\text{SE}_{Dp}}$$

where SE_{Dp} is the estimated standard deviation based on the pooled estimate

$$\hat{p} = \frac{x_1 + x_2}{n_1 + n_2}$$

of the common value $p \equiv p_1 = p_2$ of the population proportions. Here x_1 and x_2 are the number of counts possessing the characteristic being counted in sample 1 and sample 2, respectively, and

$$\text{SE}_{Dp} = \sqrt{\hat{p}(1 - \hat{p}) \left(\frac{1}{n_1} + \frac{1}{n_2} \right)}$$

The decision rules based on z are then analogous to those in Section 8.1. For example, if the alternative hypothesis is

$$H_a : p_1 > p_2$$

then we reject at level α if

$$z > z^*$$

where z^* is the upper α standard normal critical value. Furthermore, the P-value is $P(Z > z)$, where Z is $N(0, 1)$.

Example 8.4. (Adapted from Exercises 8.42 - 8.43, page 598 in the text.) In the 2000 regular baseball season, the World Series Champion New York Yankees played 80 games at home and 81 games away. They won 44 of their home games and 43 of the games played away. We can consider these games as samples from potentially large populations of games played at home and away. How much of an advantage does the Yankee home field provide?

(a) Most people think it is easier to win at home than away. Formulate null and alternative hypotheses to examine this idea.

(b) Compute the z statistic and its P-value. What conclusion do you draw?

Solution

(a) Let

$$p_1 = \text{probability that the Yankees win a home game}$$

$$p_2 = \text{probability that the Yankees win an away game}$$

Since we are trying to "prove" that it is easier to win at home than away, the appropriate significance test should be

$$H_0 : p_1 = p_2$$
$$H_a : p_1 > p_2$$

	A	B	C	D	E
1			Significance Test for the Difference in Proportions		
2					
3		Summary Statistics and User Input			
4	Group	n	X	p_hat	
5	home	80	44	0.550	
6	away	81	43	0.531	
7					
8	null	0	Calculations		
9	alpha	0.05	pooled_p	0.540	=(C5+C6)/(B5+B6)
10	alternate	upper	SE	0.079	=SQRT(pooled_p*(1−pooled_p)*(1/B5 + 1/B6))
11			z	0.244	=((D5−D6)−null)/SE
12	Lower Test				
13	lower_z		=NORMSINV(alpha)		
14	Decision		=IF(z<lower_z,"Reject HO","Do Not Reject HO")		
15	Pvalue		=NORMSDIST(z)		
16	Upper Test				
17	upper_z	1.645	=−NORMSINV(alpha)		
18	Decision	Do Not Reject HO	=IF(z>upper_z,"Reject HO","Do Not Reject HO")		
19	Pvalue	0.404	=1−NORMSDIST(z)		
20	Two-Sided Test				
21	two_z		=ABS(NORMSINV(alpha/2))		
22	Decision		=IF(ABS(z)>two_z,"Reject HO","Do Not Reject HO")		
23	Pvalue		=2*(1−NORMSDIST(ABS(z)))		

Figure 8.4: Significance Test—Difference in Proportions

(b) Fig 8.4 is a template that provides all the formulas required (located in the relevant cells in Columns C and E). Although this is an upper test, we have provided the formulas for lower and two-tailed tests.

The sample proportion of games won at home is (from cell D5)

$$\hat{p}_1 = \frac{44}{80} = 0.550$$

and the proportion of games won away is (from cell D6)

$$\hat{p}_2 = \frac{43}{81} = 0.531$$

Cell D9 gives the pooled sample proportion

$$\hat{p} = \frac{44 + 43}{80 + 81} = 0.540$$

The z test statistic in cell D11 is 0.224, which is not significant (cell B18), and the P-value in cell B19 is 0.129. There is no evidence of a difference between home and away games.

Chapter 9

Analysis of Two-Way Tables

9.1   Data Analysis for Two-Way Tables

In Example 8.3 we were interested in comparing two populations (male, female) with respect to one response variable, "frequent binge drinker." The response variable had two values, "yes" or "no." A test was carried out of the null hypothesis

$$H_0 : p_1 = p_2 \tag{9.1}$$

where p_1 and p_2 are the proportions in the respective two populations who are binge drinkers.

There is another way to view and display the data. We may consider measuring two variables, gender and binge category, on the 17,096 students who were surveyed. With two possible levels for each variable, there are four combinations of measurements, and we could display the results as a frequency table or a bar graph involving four categories. But this would cause us to lose track of how the four categories are related to the levels and the variables (essentially the "geometry" of the design). It is more natural to display the data in a different (though equivalent) way than was presented in Table 8.1 by using a "two-dimensional" frequency table, shown here in Table 9.1.

Table 9.1: Binge Drinkers—2 × 2 Table

Frequent binge drinker	Gender		Total
	Men	Women	
Yes	1630	1684	3314
No	5550	8232	13782
Total	7180	9916	17096

This presentation of the data shows that there are two variables of interest, one that might be considered as an explanatory variable (gender) and the other as the response variable. The significance test of the null hypothesis in (9.1) judges whether there is a relationship between the two variables. If $p_1 = p_2$, then there is no relationship.

We wish to generalize this test to the situation in which there are more than two populations of interest or where the response variable can take more than two values.

A table, such as Table 9.1, showing data collected on two categorical variables having r rows and c columns of values for each of the two variables is called an $r \times c$ table. In this chapter we discuss a technique based on the chi-square distribution for deciding if there is a relationship between two categorical variables.

9.2   Inference for Two-Way Tables

Example 9.1.   (Examples 9.12–9.15, pages 620–626 in the text.) Do men and women participate in sports for the same reasons? One goal for sports participants is social comparison – the desire to win or to do better than other people. Another is mastery – the desire to improve one's skills or to try one's best. A study on why students participate in sports collected data from 67 male and 67 female undergraduates at a large university. Each student was classified into one of four categories based on his or her responses to a questionnaire about sports goals. The four categories were high social comparison-high mastery (HSC-HM), high social comparison-low mastery (HSC-LM), low social comparison-high mastery (LSC-HM), and low social comparison-low mastery (LSC-LM). One purpose of the study was to compare the goals of male and female students. The data are displayed in a two-way table (Table 9.2). The entries in this table are the observed, or sample, counts. For example, there are 14 females in the high social comparison-high mastery group. Determine whether there is an association between gender and goal.

Table 9.2: Observed Counts for Sports Goals

Goal	Gender Female	Male	Total
HSC-HM	14	31	45
HSC-LM	7	18	25
LSC-HM	21	5	26
LSC-LM	25	13	38
Total	67	67	134

Example 9.1 is a direct generalization of Example 8.3. One variable, with categories displayed across the top of Table 9.1, is also gender, the explanatory variable. But now the second variable has four levels (as opposed to the two levels in Example 8.3) and the objective is to determine whether the proportions in the population for each level of the second variable are the same in males as they are in females. Another possible way of expressing this is to say that we are asking whether the cross classifications into the two variables are independent of each other. Which interpretation is appropriate depends on how the data are collected, essentially whether with fixed or random marginal totals, but the methodology is the same in either case, as discussed in the text.

Each combination of row and column levels defines a cell, and we call a table, such as Table 9.2, with (more generally) r rows and c columns an $r \times c$ contingency table. The two columns labeled "Female" and "Male" represent the results of independent samples from the respective populations.

Suppose that

$$\mathbf{p}_1 = (p_{11}, p_{12}, p_{13}, p_{14})$$
$$\mathbf{p}_2 = (p_{21}, p_{22}, p_{23}, p_{24})$$

represent the population proportions of all female (respectively male) undergraduates who are in the four goal levels. Thus, for instance, p_{12} = the population proportion among all females whose goal rating is HSC-LM, the second row in Table 9.2.

The null hypothesis of interest is

$$H_0 : \mathbf{p}_1 = \mathbf{p}_2$$

meaning equality of the vector of proportions, thereby generalizing (9.1). However, this methodology is not limited to $c = 2$, which is the example here. With c columns we would be testing

$$H_0 : \mathbf{p}_1 = \mathbf{p}_2 = \cdots = \mathbf{p}_c$$

for corresponding population proportion vectors $\mathbf{p}_i$.

Some preparation is needed before we can use the Excel function CHIINV, which returns the *P*-value for this significance test. We need to first create a table of expected cell counts.

Expected Cell Counts

Consider row 1 in Table 9.2, labeled HSC-HM, in which there are 14 counts under Female and 31 counts under Male for a total sum of row counts equal to 45.

Under the null hypothesis, we may pool the data for Male and Female with regard to the variable "Goal." Since the total number of subjects is 134, we get a pooled estimate

$$\frac{14+31}{67+67} = \frac{45}{134} = 0.3358$$

for the common value $p_{11} = p_{21}$.

The column total for "Female" is 67. Therefore the number of counts in cell "Female × HSC-HM" is a binomial random variable on 67 trials with "success probability" $\frac{45}{134}$. Under H_0 we estimate this with the usual binomial mean

$$67 \times \frac{45}{134} = \frac{45 \times 67}{134} = 22.5$$

leading to the useful mnemonic

$$\text{expected count} = \frac{\text{row total} \times \text{column total}}{\text{table total}}$$

Expected counts need to be computed for all cells. These calculations need to be hand-coded in the Excel workbook, and we now show how to do this. Copy the observed counts to an Excel workbook. This is shown in Fig 9.1 where the data appear in block A4:C8 (including labels).

	A	B	C	D	E	F	G	H
1			Computing the Expected Table					
2								
3		Observed Counts				Formulas for Expected Counts		
4		Female	Male	Row Total		Female	Male	Row Total
5	HSC-HM	14	31	45		=B$9*$D5/$D$9	=C$9*$D5/$D$9	=SUM(B14:C14)
6	HSC-LM	7	18	25		=B$9*$D6/$D$9	=C$9*$D6/$D$9	=SUM(B15:C15)
7	LSC-HM	21	5	26		=B$9*$D7/$D$9	=C$9*$D7/$D$9	=SUM(B16:C16)
8	LSC-LM	25	13	38		=B$9*$D8/$D$9	=C$9*$D8/$D$9	=SUM(B17:C17)
9	Column Total	67	67	134				
10								
11		Expected Counts						
12		Female	Male	Row Total		Chi-Square Test		
13	HSC-HM	22.5	22.5	45		P-value:	1.622E-05	
14	HSC-LM	12.5	12.5	25			=CHITEST(B5:C8,B13:C16)	
15	LSC-HM	13	13	26				
16	LSC-LM	19	19	38				
17	Column Total	67	67	134				

Figure 9.1: Expected Cell Counts and *P*-value

1. For the row totals, type the formula = SUM(B5:C5) in cell D5 and press Enter.

2. Select cell D5, then click the fill handle and fill column D to cell D8.

3. Enter the formula = SUM(B5:B8) in cell B9 and fill row 9 to cell D9 to produce the row totals and the overall table total.

Next we produce the table of expected counts. Again refer to Fig 9.1.

1. Copy the entire observed table block A3:D9 to another location, say A11:D17, and change the label "Observed Counts" to "Expected Counts."

2. Select cell B13 and enter the formula = B$9*$D5/$D$9, which involves both absolute and relative cell reference. Fill the formula to the other cells in the block B13:C16 by first filling from cell B13 to C13 and then from block B13:C13 to B16:C16. This fills the table with expected counts. We require an absolute reference $9 since the column totals are always in row 9. We require an absolute reference $D because the row totals are always in column D. We don't want these to change when the formula is filled to all the cells in the expected table, so absolute references are mandated as shown.

3. Finally obtain the marginal sums as before.

Fig 9.1 shows the formulas in block F4:H9 that are behind the cells of expected counts in block B12:D17.

The Chi-Square Test

The statistic

$$X^2 = \sum \frac{(\text{observed count} - \text{expected count})^2}{\text{expected count}}$$

where the sum is taken over counts in all the cells, was introduced by Karl Pearson in 1900 to measure how well the model H_0 fits the data. Under H_0 it has a sampling distribution that is approximated by a one-parameter family of distributions known as chi-square and denoted by the symbol χ^2. The parameter is called the degrees of freedom ν, and for an $r \times c$ contingency table it is known that $\nu = (r-1)(c-1)$. If H_0 is true, then X^2 should be "small," while if H_0 is false then X^2 should be "large." This leads to the criterion

Chi-square test: Reject H_0 if $X^2 > \chi^2$

where χ^2 is the upper critical α value of a chi-square distribution on $(r-1)(c-1)$ degrees of freedom.

The P-Value

The Excel function CHITEST calculates the P-value associated with the Pearson X^2 statistic. The syntax is

= CHITEST(actual_range, expected_range)

where "actual_range" refers to the observed table of counts B5:C8 and where "expected_range" refers to the expected table of counts B13:C16. Refer to block F13:G14 of Fig 9.1 where we have given the formula and its value. The P-value is 0.000016, so we reject H_0 and conclude that there is an association between gender and sports goals.

9.3   Formulas and Models for Two-Way Tables

Computing the Chi-Square Statistic

Although not required for the function CHITEST, it is instructive to do the calculation of X^2 and then carry out the test based on the computed value of X^2. We have shown this in block F3:G13 on the right half of Fig 9.3. For convenience the required formulas are displayed in Fig 9.2, and we present the details.

	A	B	C	D	E	F	G
1						Computing Chi Square – Formulas	
2							
3		Observed Counts				Chi-Square Cell Formulas	
4		Female	Male	Row Total		Female	Male
5	HSC-HM	14	31	45		=(B5-B13)^2/B13	=(C5-C13)^2/C13
6	HSC-LM	7	18	25		=(B6-B14)^2/B14	=(C6-C14)^2/C14
7	LSC-HM	21	5	26		=(B7-B15)^2/B15	=(C7-C15)^2/C15
8	LSC-LM	25	13	38		=(B8-B16)^2/B16	=(C8-C16)^2/C16
9	Column Total	67	67	134			
10						Chi-Square	=SUM(F5:G8)
11		Expected Counts				Critical 5% value	=CHIINV(0.05, 3)
12		Female	Male	Row Total		Decision	
13	HSC-HM	22.5	22.5	45		=IF(G10>G11, "Reject H0", "Do Not Reject H0")	
14	HSC-LM	12.5	12.5	25			
15	LSC-HM	13	13	26			
16	LSC-LM	19	19	38			
17	Column Total	67	67	134			

Figure 9.2: Computing the Value of X^2—Formulas

	A	B	C	D	E	F	G
1						Computing Chi Square – Values	
2							
3		Observed Counts				Chi-Square Cell Values	
4		Female	Male	Row Total		Female	Male
5	HSC-HM	14	31	45		3.2111	3.2111
6	HSC-LM	7	18	25		2.4200	2.4200
7	LSC-HM	21	5	26		4.9231	4.9231
8	LSC-LM	25	13	38		1.8947	1.8947
9	Column Total	67	67	134			
10						Chi-Square	24.898
11		Expected Counts				Critical 5% value	7.815
12		Female	Male	Row Total		Decision	
13	HSC-HM	22.5	22.5	45		Reject H0	
14	HSC-LM	12.5	12.5	25			
15	LSC-HM	13	13	26			
16	LSC-LM	19	19	38			
17	Column Total	67	67	134			

Figure 9.3: Computing the Value of X^2—Values

1. Copy the block B4:C8 to a convenient location, shown here copied to cells F4:G8. Change the label "Response" to "Chi-Square Cell Values."

2. The equation for X^2 is

$$X^2 = \sum \frac{(\text{observed count} - \text{expected count})^2}{\text{expected count}}$$

which we translate into Excel by the formula $= (\text{B5-B13})^2/\text{B13}$ entered in cell F5 and then filled to the block F5:G8. See Fig 9.2.

3. Sum all six cell entries by entering $= \text{SUM(F5:G8)}$ in cell G10. We get the value

$$X^2 = 24.898$$

The critical 5% χ^2 value is given by $= \text{CHIINV}(0.05, 3)$ in cell G11 and is

$$\chi^2_{.05} = 7.815$$

We therefore reject H_0, which is the same conclusion drawn using P-value.

You may arrange for the decision to appear on your workbook using the formula $= \text{IF}(G10 > G11, \text{"Reject H0"}, \text{"Do Not Reject H0"})$, which we have entered in cell G13.

Chapter 10

Inference for Regression

We pointed out in Chapter 2 that there are three procedures in Excel for regression analysis. These are complementary for the most part. We have already used the **Insert Trendline** command to fit and draw the least-squares regression line. In this chapter we derive additional information about the regression model with the **Regression** tool in the ToolPak. We also introduce some relevant Excel functions.

10.1 Simple Linear Regression

The general regression model for n pairs of observation (x_i, y_i) is

$$\mathtt{DATA} = \mathtt{FIT} + \mathtt{RESIDUAL}$$

which is expressed mathematically as

$$y_i = \beta_0 + \beta_1 x_i + \varepsilon_i$$

The function

$$\mu_y = \beta_0 + \beta_1 x$$

is called the population (true) regression curve of y on x, here taken to be linear. It represents the mean response of y as a function of x. The quantities $\{\varepsilon_i\}$ are assumed to be independent normally distributed random variables with a common mean 0 and common standard deviation σ. Therefore there are three unknown parameters $(\beta_0, \beta_1, \sigma)$.

The parameters β_0, β_1 are estimated by the method of least-squares (described in Chapter 2) by the values b_0, b_1, respectively, where

$$
\begin{aligned}
b_0 &= \bar{y} - b_1 \bar{x} \\
b_1 &= r \, \frac{s_y}{s_x}
\end{aligned}
$$

Recall that s_x and s_y are the sample standard deviations of the x and y data sets respectively.

The least-squares regression line, which estimates the population regression line, is given by the equation

$$\hat{y} = b_0 + b_1 x$$

Algebra shows that b_0 and b_1 are unbiased estimators of β_0 and β_1. They are random variables having sampling distributions:

(i) b_0 is normal with

$$\text{mean} \quad = \quad \mu_{b_0} = \beta_0$$

$$\text{standard deviation} \quad = \quad \sigma_{b_0} = \sigma\sqrt{\frac{1}{n} + \frac{\bar{x}^2}{\sum_{i=1}^{n}(x_i - \bar{x})^2}}$$

(ii) b_1 is normal with

$$\text{mean} \quad = \quad \mu_{b_1} = \beta_1$$

$$\text{standard deviation} \quad = \quad \sigma_{b_1} = \frac{\sigma}{\sqrt{\sum_{i=1}^{n}(x_i - \bar{x})^2}}$$

(iii) In a manner entirely analogous to the case of estimating the mean of a data set, we obtain an estimate of σ^2 for inference purposes. This estimate is provided by

$$s^2 = \frac{\sum_{i=1}^{n} e_i^2}{n-2}$$

where

$$e_i = y_i - \hat{y}_i$$

is called the residual at x_i and is an estimate of the sampling error $\{\varepsilon_i\}$. The denominator $(n-2)$ gives an unbiased estimate because

$$\frac{(n-2)s^2}{\sigma^2} \quad \text{is chi-squared on } (n-2) \ df$$

Regression analysis thus consists of estimating β_0, β_1, and σ^2, and making inferences about them.

The Regression Tool

Example 10.1. (Examples 10.1–10.5, pages 660–675 in the text.) Fig 10.1 shows data from a sample of 92 males aged 20 to 29 relating skinfold thickness to body density. Body density is a proxy for fat content, a variable having medical importance. In practice, fat content is found by measuring body density, the weight per unit volume of the body. High fat content corresponds to low body density. Body density is hard to measure directly – the standard method requires that

	A	B	C	D	E	F	G	H	I	J	K	L	M	N	O
1								Complete Data Set for Example 10.1							
2															
3	LSKIN	1.27	1.56	1.45	1.52	1.51	1.51	1.5	1.62	1.5	1.75	1.43	1.81	1.6	
4	DEN	1.093	1.063	1.078	1.056	1.073	1.071	1.076	1.047	1.089	1.053	1.057	1.051	1.074	
5															
6	LSKIN	1.49	1.29	1.52	1.83	1.58	1.7	1.59	2.02	1.84	1.87	1.83	1.36	1.47	
7	DEN	1.07	1.081	1.064	1.037	1.06	1.065	1.058	1.042	1.045	1.026	1.046	1.063	1.067	
8															
9	LSKIN	1.84	1.46	1.74	1.52	1.82	1.53	1.54	1.61	1.22	1.44	1.18	1.35	1.52	
10	DEN	1.05	1.063	1.06	1.078	1.055	1.059	1.076	1.06	1.086	1.083	1.084	1.082	1.071	
11															
12	LSKIN	1.64	1.73	1.38	1.3	1.73	1.57	1.74	1.41	1.53	1.54	1.83	2.08	1.91	
13	DEN	1.056	1.046	1.096	1.074	1.054	1.063	1.058	1.076	1.066	1.076	1.058	1.039	1.029	
14															
15	LSKIN	1.47	1.67	1.89	1.84	1.38	1.5	1.49	1.57	1.5	1.34	1.52	1.74	1.24	
16	DEN	1.07	1.061	1.034	1.05	1.063	1.071	1.058	1.057	1.092	1.07	1.054	1.053	1.084	
17															
18	LSKIN	1.43	2.01	1.75	1.65	1.79	1.32	1.64	1.95	1.2	1.36	1	1.73	1.56	
19	DEN	1.079	1.03	1.041	1.06	1.072	1.071	1.056	1.037	1.083	1.069	1.096	1.057	1.063	
20															
21	LSKIN	1.52	1.53	1.86	1.85	1.62	1.39	1.52	1.69	1.27	1.21	1.58	1.1	1.53	1.46
22	DEN	1.08	1.068	1.05	1.043	1.056	1.086	1.06	1.047	1.082	1.093	1.059	1.093	1.067	1.073

Figure 10.1: Skinfold Thickness Data

subjects be weighed under water. For this reason, scientists have sought variables that are easier to measure and that can be used to predict body density. Research suggests that *skinfold thickness* can accurately predict body density. To measure skinfold thickness, pinch a fold of skin between calipers at four body locations to determine the thickness, and add the four thicknesses. There is a linear relationship between body density and the logarithm of the skinfold thickness measure. The explanatory variable x is the log of the sum of the skinfold measures, denoted by LSKIN, and the response variable y is body density denoted by DEN. Density is measured in 10^3 kg/m^3 and skinfold thickness in mm. The skinfolds were measured at the biceps, triceps, subscapular and suprailiac areas. Analyse the data using the regression tool.

(a) Plot the data and confirm that a straight-line fit is appropriate.

(b) Fit the least-squares regression line to the data.

(c) Find the standard error.

(d) Give 95% confidence intervals for the parameters β_1 and β_0.

(e) Test the null hypothesis $H_0 : \beta_1 = 0$.

(f) Compute r^2. Examine the residuals for any deviation from a straight line.

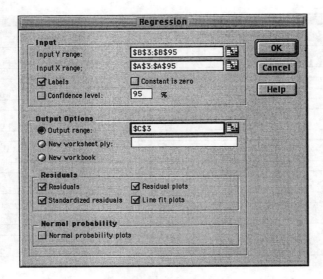

Figure 10.2: Regression Tool Dialog Box

Using the Regression Tool

1. Enter the 92 data pairs (including labels) from Fig 10.1 into columns, with the independent variable x to the left of the dependent variable y, say in columns A3:A95 and B3:B95, with the first row reserved for labels. (Excel requires this kind of data in contiguous regions.) Refer to Fig 10.3 later in this section; it shows the **Regression** tool output and also part of columns A and B.

2. From the Menu Bar choose **Tools – Data Analysis** and select **Regression** from the tools listed in the **Data Analysis** dialog box. Click OK to display the **Regression** dialog box (Fig 10.2).

3. Type the cell references B3:B95 in Fig 10.2 (or point and drag over the data in column B) for **Input Y range**. Do the same with respect to Column A for **Input X range**. Check the box **Labels**, leave **Constant is zero** clear because we are not forcing the line through the origin, and check the box **Confidence level** and insert "95." Under **Output options** select the radio button **Output range** and enter C3 to locate the upper left corner where the output will appear in the same workbook.

Check the following boxes:

Residuals.   To obtain predicted or fitted values $\hat{y}$ and their residuals.

Residual Plots.   For a scatterplot of the residuals against their x values.

Standardized Residuals. To obtain residuals divided by their standard error (useful to identify outliers).

Line Fit Plots. To obtain a scatterplot of y against x.

Do not check Normal Probability Plots.

Note: In making these selections, either use the Tab key to move from option to option or use the mouse. Click OK.

Excel Output

The output is separated into six regions: Regression Statistics, ANOVA table, statistics about the regression line parameters, residual output, scatterplot with fitted line, and residual plot. We have reproduced a portion of the output in Fig 10.3. (Part of the output also appears in the text at the top of Fig 10.5 page 669.) We next interpret the output in each of these regions.

	A	B	C	D	E	F	G	H	I
1	Regression Tool Example – Skinfold Thickness								
2									
3	LSKIN	DEN	SUMMARY OUTPUT						
4	1.27	1.093							
5	1.56	1.063	*Regression Statistics*						
6	1.45	1.078	Multiple R	0.8488					
7	1.52	1.056	R Square	0.7204					
8	1.51	1.073	Adjusted R Square	0.7173					
9	1.51	1.071	Standard Error	0.0085					
10	1.50	1.076	Observations	92					
11	1.62	1.047							
12	1.50	1.089	ANOVA						
13	1.75	1.053		*df*	*SS*	*MS*	*F*	*Significance F*	
14	1.43	1.057	Regression	1	0.01691	0.0169	231.89	1.22E-26	
15	1.81	1.051	Residual	90	0.00656	0.0001			
16	1.60	1.074	Total	91	0.02347				
17	1.49	1.070							
18	1.29	1.081		*Coefficients*	*Standard Error*	*t Stat*	*P-value*	*Lower 95%*	*Upper 95%*
19	1.52	1.064	Intercept	1.1630	0.0066	177.2962	0.0000	1.1500	1.1760
20	1.83	1.037	LSKIN	-0.0631	0.0041	-15.2281	0.0000	-0.0714	-0.0549

Figure 10.3: Data and Regression Tool Output

Parameter Estimates and Inference

Rows 19 and 20 of the output provide statistics for the regression line parameters. From cell D19 we read $b_0 = 1.163$, and from cell D20 we read $b_1 = -0.063$. The regression line is therefore

$$\hat{y} = 1.163 - 0.063x$$

Significance Tests

The test

$$H_0 : \beta_1 = 0 \qquad \text{vs.} \qquad H_a : \beta_1 \neq 0$$

is useful in assessing whether a simple model of a straight line through the origin provides an equally good fit. The appropriate test statistic

$$t = \frac{b_1}{\text{SE}_{b_1}}$$

where

$$\text{SE}_{b_1} = \frac{s}{\sqrt{\sum_{i=1}^{n}(x_i - \bar{x})^2}}$$

is the standard error of b_1 which is obtained from the standard deviation of b_1 by replacing the unknown σ with s. From E20 in Fig 10.3 we read $\text{SE}_{b_1} = 0.0041$, and from F20 we have the t statistic, denoted as t stat,

$$t = \frac{b_1}{\text{SE}_{b_1}} = \frac{-0.0631}{0.0041} = -15.228.$$

The corresponding two sided P-value appears in cell G20,

$$P\text{-value} = 0.0000$$

Similarly, for testing

$$H_0 : \beta_0 = 0 \qquad \text{vs.} \qquad H_a : \beta_0 \neq 0$$

the appropriate test statistic is

$$t = \frac{b_0}{\text{SE}_{b_0}}$$

where

$$\text{SE}_{b_0} = s\sqrt{\frac{1}{n} + \frac{\bar{x}^2}{\sum_{i=1}^{n}(x_i - \bar{x})^2}}$$

estimates the true standard deviation σ_{b_0} of b_0. From D19:G19 we read the relevant estimates and the computed test statistic

$$t = \frac{b_0}{\text{SE}_{b_0}} = \frac{1.1630}{0.0066} = 177.296$$

together with the two-sided

$$P\text{-value} = 0.0000$$

Significance tests for other null values can be carried out using the templates developed in Chapter 7, which required only the summary statistics. The main change is to use $n - 2$ instead of $n - 1$ for the degrees of freedom.

Confidence Intervals

We used the default of 95% for the confidence level when we completed the **Regression** tool dialog box. The lower and upper 95% confidence limits appear in H19:I20 in Fig 10.3. Thus 95% confidence intervals are

$$\text{for } \beta_0 \quad (1.150, 1.176)$$
$$\text{for } \beta_1 \quad (-0.0714, -0.0549)$$

Scatterplot

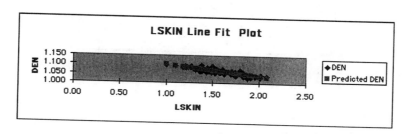

Figure 10.4: Default Scatterplot

Fig 10.4 shows one of the scatterplots produced, DEN against LSKIN. By default Excel uses markers even for the predicted values and the scales need to be changed to produce a more useful graph.

Changing Markers

We have modified the Excel scatterplot by enlarging it to make it more readable, changing the scale of the X and Y axes, replacing the diamond-shaped data plotting markers with circles, and replacing the square-shaped predicted values with a line.

1. To resize the **Chart Area** activate the Chart and drag one or more of the handles to the desired size.

2. To resize the **Plot Area** click within the plot area and drag one or more handles to the desired size.

3. To change the scale on the X axis activate the Chart and double-click the X axis. (Equivalently, click the X axis once and choose **Format − Selected Axis...** from the Menu Bar.) Then under the **Scale** tab change **Minimum** to 1 and **Maximum** to 2.1. Similarly edit the Y-axis and under the **Scale** tab set **Minimum** to 1.02 and **Maximum** to 1.11.

4. To change the data markers first activate the Chart and select one of the data points. From the Menu Bar choose **Format − Selected Data Series...**, click the **Patterns** tab, and select a **Custom Marker** (Fig 10.5).

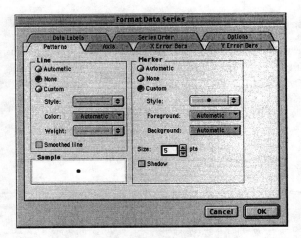

Figure 10.5: Formatting Markers

Changing Predicted Markers to a Line

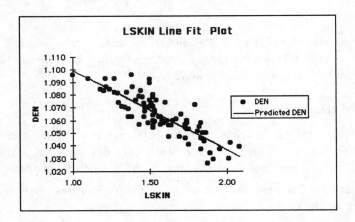

Figure 10.6: Regression Line and Enhanced Scatterplot

The other important enhancement to the markers is to change the predicted ones in the default scatterplot into a line. This is a useful enhancement even in other contexts and we separate its description here. The steps are as follows.

1. Activate the Chart and select one of the predicted markers. Choose **Format-Selected Data Series...** from the Menu Bar.

2. In the **Format Data Series** dialog box, click the **Patterns** tab. Select radio button **Custom** for **Line**, and then pick a Color. Finally, select **None**

for **Marker**. The markers disappear and are replaced by the regression line (Fig 10.6).

Residual Plot

The enhanced residual plot appears in Fig 10.7. The residuals show a random scatter about the X axis indicating that the straight-line fit is appropriate.

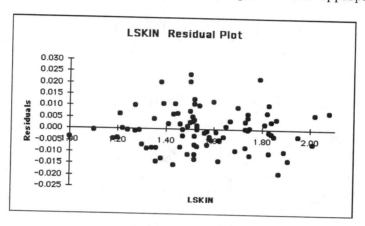

Figure 10.7: Residual Plot

	C	D	E	F
24	RESIDUAL OUTPUT			
25				
26	*Observation*	*Predicted DEN*	*Residuals*	*Standard Residuals*
27	1	1.082837953	0.010162047	1.196664361
28	2	1.064533444	-0.001533444	-0.180575620
29	3	1.071476534	0.006523466	0.768191630
30	4	1.067058204	-0.011058204	-1.302194137
31	5	1.067689394	0.005310606	0.625367386
32	6	1.067689394	0.003310606	0.389850996
33	7	1.068320584	0.007679416	0.904314182
34	8	1.060746305	-0.013746305	-1.618740007
35	9	1.068320584	0.020679416	2.435170714
36	10	1.052540835	0.000459165	0.054070411
37	11	1.072738913	-0.015738913	-1.853386038
38	12	1.048753696	0.002246304	0.264520751
39	13	1.062008684	0.011991316	1.412075676
40	14	1.068951774	0.001048226	0.123437225
41	15	1.081575573	-0.000575573	-0.067778400
42	16	1.067058204	-0.003058204	-0.360128579
43	17	1.047491316	-0.010491316	-1.235438399
44	18	1.063271064	-0.003271064	-0.385194628

Figure 10.8: Residual Output

Residual Output

Fig 10.8 shows a portion of the residual output with all predicted values, residuals and standardized residuals. With this at hand, other diagnostic scatterplots can be quickly obtained, in addition to the default output—for example, residuals versus x-variable.

10.2   More Detail about Simple Linear Regression

In this section we consider the remaining two elements of the output from the **Regression** tool, as well as inference about predicted values and use of Excel regression functions

ANOVA F and Regression Statistics

ANOVA Table

The ANOVA approach uses an F-test to determine whether a substantially better fit is obtained by the regression model than by a model with $\beta_1 = 0$. The ANOVA output breaks the observed total variation in the data

$$\text{SST} = \sum_{i=1}^{n}(y_i - \bar{y})^2$$

into two components, residual or error sum of squares

$$\text{SSE} = \sum_{i=1}^{n}(y_i - \hat{y}_i)^2$$

and a model sum of squares

$$\text{SSM} = \sum_{i=1}^{n}(\hat{y}_i - \bar{y})^2$$

connected by the identity

$$\text{SST} = \text{SSE} + \text{SSM}$$

The test criterion is based on how much smaller the residual sum of squares is under each fit and is based on the F ratio

$$F = \frac{\text{MSM}}{\text{MSE}}$$

where $\text{MSM} = \frac{\text{SSM}}{1}$ is the mean square for the model fit while $\text{MSE} = \frac{\text{SSE}}{n-2}$ is the mean square for error. Under the null hypothesis that $\beta_1 = 0$, this ratio has an F distribution with 1 degree of freedom in the numerator and $n - 2$ degrees of freedom in the denominator.

	C	D	E	F	G	H
12	ANOVA					
13		df	SS	MS	F	Significance F
14	Regression	1	0.01691	0.0169	231.89	1.22E-26
15	Residual	90	0.00656	0.0001		
16	Total	91	0.02347			

Figure 10.9: ANOVA Output

All the above calculations appear in rows 13–16 in Fig 10.9. This is an advanced topic, and the reader is referred to the text for a more complete discussion. The only point we add here is that the F test is identical to the earlier two-sided test for $H_0 : \beta_1 = 0$ vs. $H_a : \beta_1 \neq 0$ and the F statistic 231.894 in cell G14 is always the square of the t stat $= -15.228$ appearing in cell F20 in Fig 10.3 in this context.

Regression Statistics

	C	D
5	*Regression Statistics*	
6	Multiple R	0.8488
7	R Square	0.7204
8	Adjusted R Squar	0.7173
9	Standard Error	0.0085
10	Observations	92

Figure 10.10: Regression Statistics Output

This is the last component of the output from Fig 10.3 which we discuss briefly and isolate in Fig 10.10, which gives, for instance, the coefficient of determination $r^2 = 0.720$ and the standard error $s = 0.00854$, in addition to more advanced statistics such as the Adjusted R Square used in multiple regression.

Inference about Predictions

In the preceding section we drew inferences on the parameters β_0 and β_1 in the regression line $\mu_y = \beta_0 + \beta_1 x$. Here we examine the mean response and the prediction of a single outcome, both at a specified value x^* of the explanatory variable.

- To estimate the mean response, we use a confidence interval for the parameter μ_y based on the point estimate

$$\hat{\mu}_y \equiv \hat{y} = b_0 + b_1 x$$

A level C confidence interval for μ_y is given by

$$\hat{y} \pm t^* \text{SE}_{\hat{\mu}} \tag{10.1}$$

where the standard error is given by

$$\text{SE}_{\hat{\mu}} = s\sqrt{\frac{1}{n} + \frac{(x^* - \bar{x})^2}{\sum_{i=1}^{n}(x_i - \bar{x})^2}}$$

• To predict a single observation at x^*, a level C prediction interval given by

$$\hat{y} \pm t^*\text{SE}_{\hat{y}} \qquad\qquad (10.2)$$

where the appropriate standard error is now given by

$$\text{SE}_{\hat{y}} = s\sqrt{1 + \frac{1}{n} + \frac{(x^* - \bar{x})^2}{\sum_{i=1}^{n}(x_i - \bar{x})^2}}$$

In each case t^* is the upper $(1 - C)/2$ critical value of the Student t-distribution with $n - 2$ degrees of freedom.

Unfortunately Excel's Regression tool does not provide either of these intervals and (10.1) and (10.2) need to be constructed using Excel functions and appropriate cell references.

There are several ways to proceed. The Regression tool gives, inter alia, b_0, b_1, and s from which $\hat{y}$ and $\text{SE}_{\hat{\mu}}$ can be determined. We illustrate with Example 10.2. An alternative to the Regression tool uses Excel regression functions. This is the third method mentioned in Chapter 2 for regression analysis and is taken up in detail at the end of this section.

> **Example 10.2.** (Based on Examples 10.8–10.14, pages 682–687 in the text.) As in many other businesses, technological advances and new methods have produced dramatic results in agriculture. Cells A3:B7 in Fig 10.11 give the data on the average yield in bushels per acre of U.S. corn for selected years. Here, year is the explanatory variable x, and yield is the response variable y. A scatterplot suggests that we can use linear regression to model the relationship between yield and time.
>
> (a) Assuming that the linear pattern of increasing corn yield continues into the future, estimate the yield in the year 2006.
>
> (b) Give a 95% prediction interval for the yield in the year 2006 which quantifies the uncertainty in this estimate.
>
> (c) Give a 95% confidence interval for the mean yield in the year 1990.

Solution

1. Enter the four pairs of observations from the table in Example 10.8 on page 686 in the text into columns A and B of a workbook (Fig 10.11).

	A	B	C	D	E	K
1			**Confidence and Prediction Intervals**			
2						
3	Year	Yield	SUMMARY OUTPUT			
4	1966	73.1				
5	1976	88.0	*Regression Statistics*			
6	1986	119.4	Multiple R	0.976		
7	1996	127.1	R Square	0.952		
8			Adjusted R Square	0.928		
9			Standard Error	6.847		
10			Observations	4		
11						
12			ANOVA			
13						
14				*df*	*SS*	
15			Regression	1	1870.18	
16			Residual	2	93.76	
17			Total	3	1963.94	
18						
19				*Coefficients*	*Standard Error*	
20			Intercept	-3729.35	606.60	
21			Year	1.93	0.31	
22		Prediction Interval for a Future Observation				
23	Fit	150.25				
24	StDev Fit	10.83	=D9*SQRT(1+1/4+(2006-AVERAGE(A4:A7))^2/(4*VARP(A4:A7)))			
25	PI lower limit	103.67	=B23-TINV(0.05,2)*B24			
26	PI upper limit	196.83	=B23+TINV(0.05,2)*B24			
27						
28	Confidence Interval for Mean Response					
29	Fit	119.31	=D19+D20*1990			
30	StDev Fit	4.39	=D9*SQRT(1/4+(1990AVERAGE(A4:A7)^2/(4*VARP(A4:A7)))			
31	CI lower limit	100.40	=B29-TINV(0.05,2)*B30			
32	CI upper limit	138.22	=B28+TINV(0.05,2)*B30			

Figure 10.11: Confidence and Prediction Intervals

2. Following the instructions given in Example 10.1, invoke the **Regression** tool, completing the dialog box without checking the boxes for residuals, and output the results beginning in cell C3. Fig 10.11 also shows the output. The value of b_0 is in cell D19, b_1 is in D20, and s is in D9. The regression line is (also at top of page 688 in the text)

$$\hat{y} = -3729.35 + 1.93x$$

3. In a blank cell, say B23 in Fig 10.11, type the formula

$$= D19 + D20 * 2006$$

giving a predicted value $\hat{y} = 150.25$ (top of page 691 in the text). In cell B24 enter the formula

$$= D9 * \texttt{SQRT}(1 + 1/4 + (2006 - \texttt{AVERAGE}(A4:A7))\hat{\;}2/(4 * \texttt{VARP}(A4:A7)))$$

giving $\text{SE}_{\hat{y}} = 10.83$. In B25, enter the formula

$$= B23 - \texttt{TINV}(0.05, 2) * B24$$

and in B26, enter the formula

$$= B23 + \text{TINV}(0.05, 2) * B24$$

(For convenience we have shown these formulas on the workbook.) You can read the lower and upper 95% prediction interval (103.67, 196.83) in B25:B26 (as well as in the middle of page 691 in the text).

4. Entirely analogous steps are used to derive a confidence interval for the mean yield in 1990, the only difference being in the formula for the standard error SE_μ. Again, all formulas are shown in Fig 10.11. The required 95% confidence interval is (100.40, 138.22).

Regression Functions

FORECAST

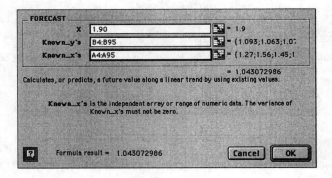

Figure 10.12: The FORECAST Dialog Box

A third method for regression analysis uses Excel functions. Suppose we wish to predict body density when the skinfold thickness explanatory variable is 1.90. Without having to derive the regression line and calculating

$$\hat{y} = 1.163 - 0.0631(1.90) = 1.043$$

we could use the FORECAST function, whose syntax is

$$\text{FORECAST}(x, \text{Known\_}y\text{'s}, \text{Known\_}x\text{'s}).$$

Here "Known_x's" refers to the set $\{x_i\}$ while "Known_y's" refers to the $\{y_i\}$. The function can be called from the **Formula Palette** or the **Function Wizard**, in which case you should select Statistical for Function category, FORECAST for Function name, and insert the required parameters in the dialog box to obtain the predicted value (Fig 10.12).

TREND

A more general method for obtaining predicted values is TREND with syntax

$$\text{TREND}(\text{Known\_}y\text{'s}, \text{Known\_}x\text{'s}, \text{New\_}x\text{'s}, \text{Const})$$

The parameter "New_x's" is the range of x-values for which predictions are desired. The parameter "Const" determines whether the regression line is forced through the origin ($\beta_0 = 0$). Use the value 1 for general β_0.

Example 10.3. (Example 10.2 continued.) Predict corn yields in the years 2000, 2001, 2002, 2003, and 2004.

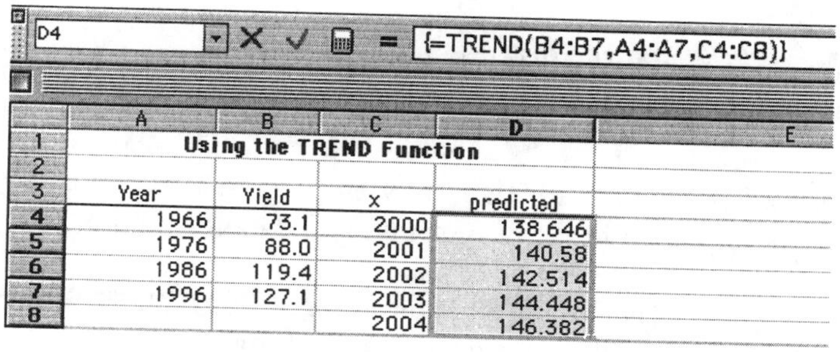

	A	B	C	D	E
1		Using the TREND Function			
2					
3	Year	Yield	x	predicted	
4	1966	73.1	2000	138.646	
5	1976	88.0	2001	140.58	
6	1986	119.4	2002	142.514	
7	1996	127.1	2003	144.448	
8			2004	146.382	

D4 = {=TREND(B4:B7,A4:A7,C4:C8)}

Figure 10.13: The TREND Function

Solution

1. Select a region of cells D4:D8 (Fig 10.13) for the output.

2. Click the **Formula Palette**, select the TREND function, and enter the values B4:B7, A4:A7, and C4:C8 into the argument fields "Known_y's," "Known_x's," and "New_x's," respectively. Leave the "Const" field blank. Click OK.

3. Click the mouse pointer in the **Formula Bar**, hold down the **Shift** and **Control** keys (either **Macintosh** or **Windows**), and press Enter to **Array-Enter** the formula. The formula will appear surrounded by braces in the Formula Bar, indicating that it has been array-entered, and the predicted values {138.648, 140.580, 142.514, 144.448, 146.382} will appear in the output range (Fig 10.13).

LINEST

The last regression function described here is LINEST, which returns the estimated least-squares line coefficients, their standard errors, r^2, s, the computed F statistic with its degrees of freedom, and the regression and error sums of squares. The syntax is

$$\text{LINEST}(x, \text{Known\_}y\text{'s}, \text{known\_}x\text{'s}, \text{Const}, \text{Stats})$$

	A	B	C	D	E
1		**Using the LINEST Function**			
2					
3	Year	Yield			
4	1966	73.1	1.934	-3729.354	
5	1976	88.0	0.30621	606.603376	
6	1986	119.4	0.95226	6.84697013	
7	1996	127.1	39.892	2	
8			1870.18	93.762	
9					
10			b1	b0	
11			SE_b1	SE_b0	
12			R Square	s	
13			F	df	
14			SSM	SSE	

Figure 10.14: The LINEST Function

If the field "Stats" is set to false, then only the regression coefficients are output; true will produce an entire block of output. The default is false.

> **Example 10.4.** (Example 10.2 continued.) Use LINEST to perform a regression analysis of corn yield versus time for the data in Fig 10.11.

Solution

1. Select a block of cells C4:D8 with two columns and five rows. (For multiple regression you require a block in which the number of columns equals the number of independent variables plus one.)

2. Click the **Formula Palette**, select the LINEST function, and enter the values B4:B7 and A4:A7 into the argument fields "Known_y's" and "Known_x's," respectively. Leave the "Const" field blank and type true in the "Stats" field. Click OK.

3. The formula = LINEST(B4:B7,A4:A7) will be visible in the Formula Bar, but only the value 1.93400 in cell C4 will appear in your sheet. With the block C4:D8 still selected, click the mouse pointer in the **Formula Bar**, as you did for the TREND function, hold down the **Shift** and **Control** keys, and press Enter to **Array-Enter** the formula. The formula will appear surrounded by braces in the Formula Bar, indicating that it has been array-entered, and the output will appear in C4:C8 (Fig 10.14)

In cells C10:D14 in Fig 10.14, we have described the contents of the values in C4:D8. These may be compared with the output from the **Regression** tool for this data set, shown in Fig 10.11. For instance, cell D8 in Fig 10.14 contains the value 93.76200. This represents the error sum of squares SSE, as seen from the corresponding description in D14, which is identical to the output in cell E15 of the **Regression** tool output in Fig 10.11.

Note: Both TREND and LINEST can be used with multiple regression.

Chapter 11

Multiple Regression

Multiple regression extends simple linear regression by fitting surfaces to data involving two or more explanatory variables $\{x_1, \ldots, x_p\}$ used to predict a response variable y. For example,

We want to predict the college grade point average of newly admitted students. We have data on their high school grades in several subjects and their scores on the two parts of the Scholastic Aptitude Test (SAT). How well can we predict college grades from this information? Do high school grades or SAT scores predict college grades more accurately?

11.1 Inference for Multiple Regression

There are n sets of observations

$$(x_{i1}, x_{i2}, \ldots, x_{ip}, y_i) \qquad 1 \leq i \leq n$$

where y_i is the response and $(x_{i1}, \ldots, x_{ip})$ and the values of the p explanatory variables measured on the ith subject.

As in Chapter 10, the model is

$$\text{DATA} = \text{FIT} + \text{RESIDUAL}$$

expressed mathematically as

$$y_i = \beta_0 + \beta_1 x_{i1} + \beta_2 x_{i2} + \cdots + \beta_p x_{ip} + \varepsilon_i$$

The FIT portion involving the $\{\beta_i\}$ is a linear model and expresses the population regression equation

$$\mu_y = \beta_0 + \beta_1 x_1 + \cdots + \beta_p x_p$$

which is the mean of the response variable for explanatory variables $(x_1, \ldots, x_p)$. The RESIDUAL component represents the variation in the observations and the

$\{\varepsilon_i\}$ are assumed to be independent and identically distributed $N(0, \sigma)$ random variables.

The regression coefficient β_i measures how much the response changes if x_i changes one unit, keeping all other explanatory variables fixed.

The parameters are estimated by the Principle of Least-Squares described in Section 10.1, and the estimates of $\{\beta_1, \ldots, \beta_p\}$ are denoted by

$$\{b_1, \ldots, b_p\}$$

giving a predicted response of

$$\hat{y}_i = b_0 + b_1 x_{i1} + b_2 x_{i2} + \cdots + b_p x_{ip}$$

The ith residual is

$$e_i = y_i - \hat{y}_i$$

and is used to estimate the population variance using

$$s^2 = \frac{\sum_{i=1}^n e_i^2}{n - p - 1}$$

Excel uses the **Regression** tool for multiple regression. The data is entered into a workbook one row for each observation so that the columns correspond to the explanatory variables, followed by the response variable (all in adjacent columns).

As in Chapter 10, the output is separated into six regions: regression statistics, ANOVA table, statistics about parameters, residuals, scatterplots, and residual plots.

The statistics about the parameters allow you to

- construct confidence intervals of the form

$$b_j \pm t^* \mathrm{SE}_{b_j}$$

  for the parameter β_j where SE_{b_j} is the standard error of b_j

- test the null hypothesis

$$H_0 : \beta_j = 0$$

  using the test statistic

$$t = \frac{b_j}{\mathrm{SE}_{b_j}}$$

The ANOVA table provides an F statistic for testing

$$H_0 : \beta_1 = \beta_2 = \cdots = \beta_p = 0$$
$$H_a : \text{ at least one of the } \beta_j \text{ is not } 0$$

and gives the estimate s^2.

The various plots can be used for diagnostic purposes and the regression statistics include the squared multiple correlation coefficient.

11.2 A Case Study

Example 11.1. (Example 11.1, page 710 and continued on pages 717 – 721 of the text. Complete data appear in the CSDATA set in the Student CD-ROM.) Data was collected at a large university on all first-year computer science majors in a particular year. The purpose of the study was to attempt to predict success in early university years. One measure of success was the cumulative grade point average (GPA) after three semesters. Among the explanatory variables recorded at the time the students enrolled in the university were average high school grades in mathematics (HSM), science (HSS), and English (HSE), coded on a scale from 1 to 10. Use the 3 explanatory variables, x_1 = HSM, x_2 = HSS, and x_3 = HSE to predict the response variable, the cumulative grade point average, in a multiple regression model

$$\mu_{GPA} = \beta_0 + \beta_1 \text{ HSM } + \beta_2 \text{HSS } + \beta_3 \text{ HSE}$$

Also recorded were SATM (SAT Math score) and SATV (SAT Verbal score). The first 20 cases in the full data set (in cells A3:G227) are shown in Fig 11.1

	A	B	C	D	E	F	G
1	\multicolumn{7}{c	}{Multiple Regression Case Study}
2							
3	Student	HSM	HSS	HSE	SATM	SATV	GPA
4	1	10	10	10	670	600	3.32
5	2	6	8	5	700	640	2.26
6	3	8	6	8	640	530	2.35
7	4	9	10	7	670	600	2.08
8	5	8	9	8	540	580	3.38
9	6	10	8	8	760	630	3.29
10	7	8	8	7	600	400	3.21
11	8	3	7	6	460	530	2.00
12	9	9	10	8	670	450	3.18
13	10	7	7	6	570	480	2.34
14	11	9	10	6	491	488	3.08
15	12	5	9	7	600	600	3.34
16	13	6	8	8	510	530	1.40
17	14	10	9	9	750	610	1.43
18	15	8	9	6	650	460	2.48
19	16	10	10	9	720	630	3.73
20	17	10	10	9	760	500	3.8
21	18	9	9	8	800	610	4.00
22	19	9	6	5	640	670	2.00
23	20	9	10	9	750	700	3.74

Figure 11.1: GPA Data Set

Preliminary Exploratory Data Analysis

In this complex setting with several explanatory variables, it is important to use some of the descriptive tools described in Chapters 1 and 2 as preliminary steps to first obtain numerical and graphical summaries. These are shown in Fig 11.2, Fig 11.3, and Fig 11.4 and are obtained as follows.

1. Referring to Fig 11.2, which is part of the same workbook as Fig 11.1, label the cells as shown in I3:N3 and H4:H7.

	H	I	J	K	L	M	N
3		HSM	HSS	HSE	SATM	SATV	GPA
4	mean	8.3214	8.0893	8.0938	595.29	504.55	2.6352
5	Std_Dev	1.6387	1.6997	1.5079	86.40	92.61	0.7794
6	minimum	2	3	3	300	285	0.12
7	maximum	10	10	10	800	760	4

Figure 11.2: Summary Statistics

2. Select H4 and enter = **AVERAGE**(B4:B227). In our workbook, B4:B227 is the range for all the HSM scores. Use whatever range is appropriate in your own workbook. In I5 enter = **STDEV**(B4:B227), in I6 enter = **MIN**(B4:B227), and in I7 enter = **MAX**(B4:B227).

3. Select I4:I7, click the fill handle in the lower right corner of I7, and fill to the range I4:N7.

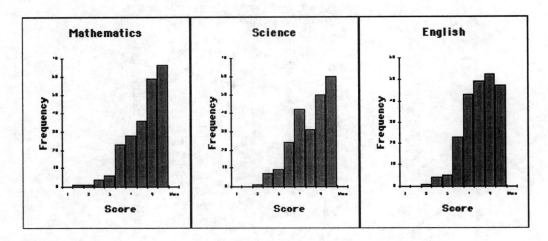

Figure 11.3: Histograms

Next, we choose **Tools – Data Analysis – Histogram** from the Menu Bar to produce histograms for the explanatory variables (refer to Chapter 1). Fig 11.3 shows histogram for the three explanatory variables scores.

We can also examine pairwise correlations for all variables by choosing **Tools – Data Analysis – Correlation** from the Menu Bar, as in Section 2.2. The correlation matrix appears in Fig 11.4.

	H	I	J	K	L	M	N
28		*HSM*	*HSS*	*HSE*	*SATM*	*SATV*	*GPA*
29	HSM	1					
30	HSS	0.5757	1				
31	HSE	0.4469	0.5794	1			
32	SATM	0.4535	0.2405	0.1083	1		
33	SATV	0.2211	0.2617	0.2437	0.4639	1	
34	GPA	0.4365	0.3294	0.2890	0.2517	0.1145	1

Figure 11.4: Pairwise Correlations

Estimation of Parameters and ANOVA Table

Using the Regression Tool

Having explored some distributional aspects of the explanatory and response variables, we are ready to run a multiple regression. We use the same dialog box as in the one-variable regression that appears after we choose **Tools – Data Analysis – Regression** from the Menu Bar, so we will not repeat the explicit steps given in Chapter 10 but refer the reader to that chapter.

Excel Output

Fig 11.5 shows a portion of the multiple regression output for a model involving only the three predictor variables HSM, HSS, HSE. The ANOVA F statistic is given in cell W12 as 18.861. Cells T12:T13 show that the numerator degrees of freedom are DFM $= p = 3$ and the error (residual) degrees of freedom are DFE $= n - p - 1 = 224 - 3 - 1 = 220$. The P-value is 0.0000 in cell X12, and therefore we reject

$$H_0 : \beta_1 = \beta_2 = \beta_3 = 0$$

and conclude that at least one of the three regression coefficients is not 0. Cell T7 is the estimate for standard error

$$s = 0.700$$

and in cell T5 we see that the squared multiple correlation is

$$R^2 = 0.205$$

which indicates that 20.5% of the observed variation in the GPA scores $\{y_i\}$ is accounted for by the linear regression on these three high school scores. At the bottom of Fig 11.5, we find in cells T17:T20 the least-squares estimates for $\beta_0, \beta_1, \beta_2$, and β_3, respectively. Thus the fitted regression equation is

$$\widehat{\text{GPA}} = 0.590 + 0.169 \text{ HSM} + 0.034 \text{ HSS} + 0.045 \text{ HSE}$$

Compare with the same equation shown in the middle of page 721 and on page 725 in the text.

In the same portion of the output we can find confidence intervals (X17:Y20) and values of the t-test statistics (V17:V20) as well as P-values (W17:W20). We conclude that only HSM is a significant explanatory variable.

	S	T	U	V	W	X	Y
1	SUMMARY OUTPUT						
2							
3	*Regression Statistics*						
4	Multiple R	0.452					
5	R Square	0.205					
6	Adjusted R Square	0.194					
7	Standard Error	0.700					
8	Observations	224					
9							
10	ANOVA						
11		df	SS	MS	F	Significance F	
12	Regression	3	27.712	9.237	18.861	0.0000	
13	Residual	220	107.750	0.490			
14	Total	223	135.463				
15							
16		Coefficients	Standard Error	t Stat	P-value	Lower 95%	Upper 95%
17	Intercept	0.5899	0.2942	2.0047	0.0462	0.0100	1.1698
18	HSM	0.1686	0.0355	4.7494	0.0000	0.0986	0.2385
19	HSS	0.0343	0.0376	0.9136	0.3619	-0.0397	0.1083
20	HSE	0.0451	0.0387	1.1655	0.2451	-0.0312	0.1214

Figure 11.5: Regression Tool Output

Interpretation

Some of the numerical results appear contradictory. The value $R^2 = 0.205$ is small, indicating that the model does not explain much of the variation. Yet the small P-value for the test $H_0 : \beta_1 = 0$ against $H_a : \beta_1 \neq 0$ suggests that the HSM score is significant.

Moreover, if we ran simple regressions of GPA against each of HSS and HSE we would find that the corresponding explanatory variables taken individually are significant, while all three taken together are not.

A partial explanation for this may be found in the relatively high correlations between HSM and HSS, HSE as shown in Fig 11.4. This means that there is overlap in predictive information contained in these variables.

Residuals

Residual plots are useful diagnostic aids for checking not merely the linearity but, for instance, the assumption of constant variance in a model. Excel provides plots of the residuals against each of the explanatory variables. These plots appear to the right of the numerical summaries and are omitted here.

Chapter 12

One-Way Analysis of Variance

Analysis of variance (ANOVA) is a technique for comparing the means of two or more populations. It is a direct generalization of the two-sample t test described in Chapter 7. In particular, the F statistic in ANOVA, when there are two populations, is precisely the square of Student's t, and the ANOVA F test is then identical to the Student two-sample t procedure.

As the name suggests, ANOVA consists of separating the variability in a data set into two components and judging whether a fit that assumes equal population means is substantially better than a fit in which all means are assumed to be the same. This is achieved by comparing the residual variation following the model fit in both cases by a ratio called an F, which may be viewed as a signal-to-noise ratio. (See also the discussion in Chapter 10 on the ANOVA output from the **Regression** tool.)

12.1 Inference for One-Way Analysis of Variance

Suppose there are I normal populations labeled $1 \leq i \leq I$ and that independent random samples $\{x_{ij} : 1 \leq j \leq n_i\}$ of size $n_i \geq 1$, $1 \leq i \leq I$ are taken from each. The ANOVA model is

$$x_{ij} = \mu_i + \varepsilon_{ij} \qquad 1 \leq i \leq I,\ 1 \leq j \leq n_i$$

where $\{\varepsilon_{ij}\}$ are independent $N(0, \sigma)$ random variables. The populations thus have means μ_i and a common standard deviation σ, and we express this as

$$\texttt{DATA} = \texttt{FIT} + \texttt{RESIDUAL}$$

The one-way ANOVA significance test is

$$H_0 : \mu_1 = \mu_2 = \cdots = \mu_I$$
$$H_a : \text{not all of the } \mu_i \text{ are equal}$$

Under H_0 we estimate the common value μ with the overall mean

$$\bar{x} = \sum_{i=1}^{I} \sum_{j=1}^{n_i} x_{ij}$$

What is left over $x_{ij} - \bar{x}$ is called the residual under H_0, and then the total residual variation in the data is given by

$$\mathrm{SST} = \sum_{i=1}^{I} \sum_{j=1}^{n_i} (x_{ij} - \bar{x})^2$$

called the **total sum of squares.**

If we do not assume H_0 is true, then we should estimate each individual μ_i by the corresponding sample mean $\bar{x}_i$. The remainder $x_{ij} - \bar{x}_i$ is then the residual under an unrestricted model, and the total residual variation in the data is given by

$$\mathrm{SSE} = \sum_{i=1}^{I} \sum_{j=1}^{n_i} (x_{ij} - \bar{x}_i)^2$$

called the **error** (or within groups) **sum of squares.** Intuitively, a better fit always occurs with the unrestricted model. We can show with a little algebra that the difference $\mathrm{SSG} = \mathrm{SST} - \mathrm{SSE}$ is positive and can also be expressed as

$$\mathrm{SSG} = \sum_{i=1}^{I} \sum_{j=1}^{n_i} (\bar{x}_i - \bar{x})^2$$

called the **between groups sum of squares.** Remarkably, it turns out that

$$\mathrm{SST} = \mathrm{SSG} + \mathrm{SSE}$$

which is the key to the partition of the variation.

The magnitude of SSG measures the improvement in the fit as measured by the residual sum of squares. In order to reject H_0, the improvement must be significantly beyond what might be expected due to chance, and one is led to consider the ratio $\frac{\mathrm{SSG}}{\mathrm{SSE}}$. Define the mean squares

$$\mathrm{MSG} = \frac{\mathrm{SSG}}{I-1} \qquad \mathrm{MSE} = \frac{\mathrm{SSE}}{N-I}$$

where $N = \sum_{i=1}^{I} n_i$, and form the ratio

$$F = \frac{\mathrm{MSG}}{\mathrm{MSE}}$$

known to have an F distribution with $I - 1$ degrees of freedom for the numerator and $N - I$ degrees of freedom for the denominator (denoted by $F(I - 1, N - I)$).

The decision rule is

$$\text{Reject } H_0 \text{ at level } \alpha \text{ if } F > F^*$$

where F^* is the upper α critical value of an $F(I-1, N-I)$ distribution, that is, F^* satisfies

$$P\left(F(I-1, N-I) > F^*\right) = \alpha$$

We observe that MSE can also be expressed as

$$\text{SSE} = \sum_{i=1}^{I}(n_i - 1)s_i^2$$

where $\{s_i\}$ are the sample variances and, consequently, MSE is a pooled sample variance

$$s_p^2 \equiv \text{MSE} = \frac{\sum_{i=1}^{I}(n_i - 1)s_i^2}{\sum_{i=1}^{I}(n_i - 1)}$$

and therefore an unbiased estimate of σ^2.

Finally, we define the ANOVA coefficient of determination

$$R^2 = \frac{\text{SSG}}{\text{SST}}$$

as the fraction of the total variance "explained by model H_0."

Testing Hypotheses in a One-Way ANOVA

We illustrate the implementation of the preceding discussion with the following worked exercise.

Example 12.1. (Exercise 12.30, page 789 in the text.) The presence of harmful insects in farm fields is detected by erecting boards covered with a sticky material and then examining the insects trapped on the board. To investigate which colors are most attractive to cereal leaf beetles, researchers placed six boards of each of four colors in a field of oats. Table 12.1 gives data on the number of cereal leaf beetles trapped in July.

(a) Make a table of means and standard deviations for the four colors, and plot the data and the means.
(b) State H_0 and H_a for an ANOVA on these data, and explain in words what ANOVA tests in this setting.
(c) Using Excel, run the ANOVA. What are the F statistic and its P-value? State the values of s_p and R^2. What do you conclude?

Table 12.1: Luminescent Colors and Insect Attractiveness

Color	Insects trapped					
Lemon yellow	45	59	48	46	38	47
White	21	12	14	17	13	17
Green	37	32	15	25	39	41
Blue	16	11	20	21	14	7

Plotting the Data and the Sample Means

The first step in ANOVA is usually exploratory, where the data and means are plotted. Such a plot will help to visually confirm or dispel the assumption of equal variances and to indicate possible outliers or skewness in the data that might call into question use of this technique.

The step is easily carried out in Excel using the **ChartWizard**, which produces side-by-side displays of the samples $\{x_{ij}\}$, their sample means $\{\bar{x}_i\}$, and the overall mean $\bar{x}$, a good preliminary display of ANOVA data.

Means and Standard Deviations

The mean and standard deviations are evaluated with the Excel function AVERAGE and STDEV. Refer to Fig 12.1 throughout.

1. Enter the data and labels in B3:E9 of a workbook.

2. Enter the label "mean" in A10 and then the formula = AVERAGE(B4:B9) in B10. The value 47.17 appears, which is the sample mean of the "Yellow" observations. Select cell B10 and, using the fill handle, drag to E10, filling the cells with the means for the other colors.

3. Enter the label "stdev" in A11 and the formula = STDEV(B4:B9) in B11. Then select B11 and drag the fill handle to E11. Now the standard deviations of all the samples appear.

Plotting

The workbook needs to be prepared for the **ChartWizard** by relocating the data, coding the samples, relocating the sample means, and entering the overall mean. Complete columns G, H, I, J as indicated.

Since the sample means have already been calculated in B10:E10, an efficient way to relocate them to I28:I31 is to select B10:E10, choose **Edit − Copy** from the Menu Bar, then select cell I28 and choose **Edit − Paste Special** from the Menu Bar. In the **Paste Special** dialog box check the the radio button for **values** because the contents of B10:E10 are formulas, not values, and a straight **Edit − Copy**

	A	B	C	D	E	F	G	H	I	J
1				One-Way Analysis of Variance						
2										
3		Yellow	White	Green	Blue		Color	Insects	Means	Overall
4		45	21	37	16		1	45		
5		59	12	32	11		1	59		
6		48	14	15	20		1	48		
7		46	17	25	21		1	46		
8		38	13	38	14		1	38		
9		47	17	41	7		1	47		
10	mean	47.17	15.67	31.33	14.83		2	21		
11	stdev	6.795	3.327	9.771	5.345		2	12		
12							2	14		
13							2	17		
14							2	13		
15							2	17		
16							3	37		
17							3	32		
18							3	15		
19							3	25		
20							3	38		
21							3	41		
22							4	16		
23							4	11		
24							4	20		
25							4	21		
26							4	14		
27							4	7		
28							1		47.17	27.25
29							2		15.67	27.25
30							3		31.33	27.25
31							4		14.83	27.25

Figure 12.1: Preparing the Data for the One-Way ANOVA Tool

will change the relative cell references. Then check the box **Transpose** to convert this row selection into a column and click OK. Finally, the common value 27.25 in J28:J31 is the overall mean obtained from the Excel formula = AVERAGE(B4:E9).

Users of Excel 5/95

Begin as always by clicking the **ChartWizard** button.

- In Step 1 of 5 enter G3:J31 for the range. This will produce a simultaneous scatterplot with all three variables—data, sample means, overall mean— plotted on the y-axis (with different markers) against the colors (column G) on the x-axis.

- In Step 2 select **XY (Scatter)**.

- In Step 3 select Format **1**.

- In Step 4 use **Columns** for Data Series in, First "1" Column for X Data, and First "1" Row for Legend Text.

- In Step 5 select **Yes** for Add a Legend?, type "Plot of the Data and the Means" for the Chart Title, type "Color" for Axis Title Category (X), and type "Insects Trapped" for Axis Title Value (Y). Click Finish.

Users of Excel 97/98/2000/2001

- In Step 1 select **XY (Scatter)** for Chart Type and the upper left Chart sub-type **Scatter**.

- In Step 2 under under the **Data Range** tab enter G3:J31 (which will produce a simultaneous scatterplot with all three variables – data, sample means, and overall mean – plotted on the y-axis (with different markers) against the colors (column G) on the x-axis, and select the Series radio button for **Columns**.

- In Step 3 under the **Titles** tab, type "Plot of the Data and the Means" as the Chart Title, "Color" as Value (X) Axis, and "Insects Trapped" as Value (Y) Axis. Under the **Axes** tab, both check boxes should be selected. Under the **Gridlines** tab clear all check boxes. Under the **Legend** tab check the **Show legend** and locate it to the right. Finally, under the **Data Labels** tab select the radio button **None**.

- In Step 4 embed the graph in the current workbook by selecting the radio button **As object in**. Click Finish.

A scatterplot like the one shown in Fig 12.2 appears with distinct markers representing the individual observations, the sample means, and the overall mean. Editing will enhance the usefulness of this plot.

Editing the Plot

We edit Fig 12.2 to more clearly emphasize the observations, the sample means, and the overall mean. The result resembles side-by-side boxplots.

1. Activate the chart for editing.

2. Click one of the markers representing a sample mean (the entire series will then appear highlighted). From the Menu Bar, choose **Format – Selected Data Series....** In the dialog box under the **Patterns** tab, select radio buttons **Automatic** for **Line** and **None** for **Marker**. Click OK. The markers for sample means are now replaced by a connecting polygonal line.

3. Repeat this step after selecting a marker for the overall mean. The four markers are replaced by a horizontal line.

4. Click a marker for the observations and change the Style of the marker to a square.

5. Select the horizontal axis, and from the Menu Bar choose **Format – Selected Axis....** In the dialog box click the **Scale** tab. Then type "1" for **Minimum**, "4" for **Maximum**, "1" for **Major unit** and "1" for **Minor unit**. Then click the **Patterns** tab, click **None** for **Tick Mark Labels**, and click **None** for **Major** and **Minor** Tick Mark Type. Click OK.

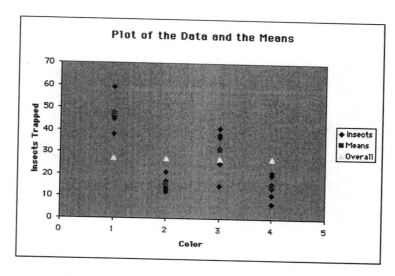

Figure 12.2: Default Plot—Data and Means

6. Activate the **Plot Area** and change the background color to white using the color swatch. Click OK.

7. Type the word "Yellow" in the **Formula Bar** and press enter. The word "Yellow" appears in a grey shaded text rectangle. With the cursor, drag it to the location shown in Fig 12.3. Click outside the text rectangle to deselect it. Now repeat for the other three labels "White," "Green," and "Blue."

The result of this enhancement is Fig 12.3, which immediately shows the dominant features of the data set much more aggressively than the numerical counterpart in cells B3:E11 of Fig 12.1.

Using the One-Way ANOVA Tool

1. From the Menu Bar choose **Tools – Data Analysis** and select **Anova: Single Factor** from the tools listed in the **Data Analysis** dialog box. Click OK to display the Anova: Single Factor dialog box (Fig 12.4).

2. Type the cell references B3:E9 (or point and drag on the workbook) for **Input range**. Select the radio button Grouped By: **Columns**. Check the box **Labels in first row** and use "0.05" for Alpha. Enter K1 for **Output range**. Click OK.

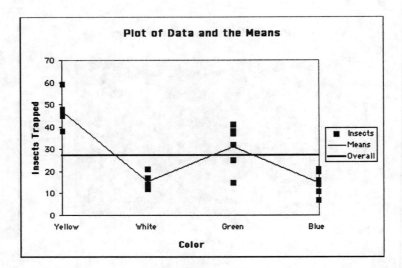

Figure 12.3: Enhanced Plot—Data and Means

Excel Output—ANOVA Table

The output from the **Anova: Single Factor** tool appears in Fig 12.5, from which we can read off all the variables described earlier in this chapter. The first half of the output provides summary statistics for the four samples. Had we not desired a plot we would have read the sample means and standard deviations directly from this output.

Source of Variation

Between Groups sum of squares refers to

$$SSG = 4210.167$$

Within Groups (error) sum of squares is

$$SSE = 906.333$$

The Total sum of squares is

$$SST = 5116.5$$

Further, we read the degrees of freedom (df)

$$DFG = 3 \quad DFE = 20 \quad DFT = 23$$

Consequently, the mean squares (MS) are

$$MSG = \frac{4210.167}{3} = 1403.389$$

$$MSE = \frac{906.333}{20} = 45.317$$

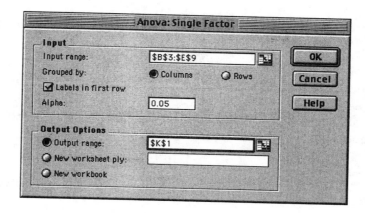

Figure 12.4: Single-Factor ANOVA Dialog Box

	K	L	M	N	O	P	Q
3	Anova : Single Factor						
4							
5	SUMMARY						
6	*Groups*	*Count*	*Sum*	*Average*	*Variance*		
7	Yellow	6	283	47.167	46.167		
8	White	6	94	15.667	11.067		
9	Green	6	188	31.333	95.467		
10	Blue	6	89	14.833	28.567		
11							
12							
13	ANOVA						
14	*Source of Variation*	*SS*	*df*	*MS*	*F*	*P-value*	*F crit*
15	Between Groups	4210.167	3	1403.389	30.97	1.03E-07	3.10E+00
16	Within Groups	906.333	20	45.317			
17							
18	Total	5116.5	23				

Figure 12.5: Single-Factor ANOVA Output

and the calculated F statistic is

$$F = \frac{\text{MSG}}{\text{MSE}} = \frac{1403.389}{45.317} = 30.97$$

The 5% critical F value (F *crit*) is

$$F^* = 3.098$$

The P-value is 0.00000.

We conclude that for the significance test

$$H_0 : \mu_1 = \mu_2 = \mu_3 = \mu_4$$
$$H_a : \text{ not all of the } \mu_i \text{ are equal}$$

we reject H_0 at the 5% level (in fact at *any* reasonable level) in view of the small P-value.

We observe that the pooled estimate of variance is

$$s_p^2 = \text{MSE} = 45.17$$

which we note is also (because of the equal sample sizes) the average of the individual sample variances.

Finally, although the coefficient of determination is not provided, we can do the arithmetic to derive

$$R^2 = \frac{\text{SSG}}{\text{SST}} = \frac{4210.167}{5116.5} = 0.823$$

Chapter 13

Two-Way Analysis of Variance

In a one-way ANOVA, independent samples are taken from I populations each differing with respect to one categorical variable, the population mean. We can view this variable as representing the levels of a particular factor as discussed in Chapter 3 of the text. In this setting the one-way ANOVA then describes the analysis of a **completely randomized design**.

Another design discussed in Chapter 3 was a **randomized block design**. We recall that a block is a group of experimental units similar in some way that is expected to influence the response. The paired two-sample t procedure is an example of a block design analysis.

Two-way ANOVA may then be viewed as a generalization of the paired t test when there are more than two populations. The approach parallels one-way ANOVA by partitioning the total variation into components that can be interpreted as representing the contributions of factor effects or block effects.

However, the analysis of a randomized block design also applies when there are two or more factors whose effect is of interest, where one is not necessarily a blocking variable and where there are multiple observations (replications) for each combination of factor levels. That is the model we consider here.

13.1 The Two-Way ANOVA Model

The data comprise samples of size n_{ij} for $I \times J$ treatments representing the combinations of I levels of factor A and J levels of factor B. The data are represented by $\{x_{ijk} : 1 \le i \le I,\ 1 \le j \le J,\ 1 \le k \le n_{ij}\}$, where x_{ijk} represents the kth observation of treatment combination (i, j). The model states

$$x_{ijk} = \mu_{ij} + \varepsilon_{ijk} \quad 1 \le I,\ 1 \le j \le J,\ 1 \le k \le n_{ij}$$

where $\{\varepsilon_{ijk}\}$ are independent $N(0, \sigma)$ random variables. Note the assumption of common variance, as with the one-way model.

First, we view the model as a one-way ANOVA for $I \times J$ populations (treatments).

$$\text{DATA} = \text{FIT} + \text{RESIDUAL}$$

In particular, we can estimate an overall mean μ by

$$\bar{x} = \frac{1}{N} \sum_{i=1}^{I} \sum_{j=1}^{J} \sum_{k=1}^{n_{ij}} x_{ijk}$$

and then produce a corresponding **total sum of squares** where N is the overall total number of observations.

$$\text{SST} = \sum_{i=1}^{I} \sum_{j=1}^{J} \sum_{k=1}^{n_{ij}} (x_{ijk} - \bar{x})^2$$

Likewise, we can estimate the residual or error variation using the sample means for each (i, j) combination of $I \times J$ treatments

$$\bar{x}_{ij} = \frac{1}{n_{ij}} \sum_{k=1}^{n_{ij}} x_{ijk}$$

and then we have the error (residual) sum of squares

$$\text{SSE} = \sum_{i=1}^{I} \sum_{j=1}^{J} \sum_{k=1}^{n_{ij}} (x_{ijk} - \bar{x}_{ij})^2$$

As with the one-way ANOVA this formula can be re-expressed as

$$\text{SSE} = \sum_{i=1}^{I} \sum_{j=1}^{J} (n_{ij} - 1) s_{ij}^2$$

where s_{ij}^2 is the sample variance for the (i, j) combination, which leads to a pooled estimate for the common variance σ^2

$$s_p^2 = \frac{\text{SSE}}{\sum_{i=1}^{I} \sum_{j=1}^{J} (n_{ij} - 1)} = \frac{\text{SSE}}{N - IJ}$$

The corresponding between groups sum of squares is here denoted as (with a change in notation, SSM replacing SSG)

$$\text{SSM} = \sum_{i=1}^{I} \sum_{j=1}^{J} \sum_{k=1}^{n_{ij}} (\bar{x}_{ij} - \bar{x})^2$$

To this stage, the analysis is for a one-way ANOVA. But because of the way the data were collected (the design), it is then possible to partition SSM further into additional sums of squares that can be identified as arising from model parameters. We therefore prescribe μ_{ij} using a **linear model**

$$\mu_{ij} = \mu + \alpha_i + \beta_j + \gamma_{ij}$$

where μ is an overall mean, α_i represents an effect due to level i of factor A, β_j represents an effect due to level j of factor B (both of these are called **main effects**), and γ_{ij} represents an **interaction** effect between factor A and factor B. We say that two factors interact if the difference in mean response for two levels of one factor is not constant across levels of the second factor.

Excel provides two versions of **two-way ANOVA** in the **Analysis ToolPak**. The first is where $n_{ij} = 1$ for all (i, j) (in which case γ_{ij} cannot be estimated). The other is where $n_{ij} \equiv n \geq 2$, so there is replication of observations at each (i, j) level but the number of replications is the same. This is called a **balanced** design. With unequal sample sizes, the ANOVA formulas become more complex and the factor effect components (sums of squares) are no longer orthogonal (they don't add up). For this reason, we now limit the discussion to balanced designs with the same number of observations $n_{ij} \equiv n$ per treatment.

The parameters in the representation of μ_{ij}

$$\mu_{ij} = \mu + \alpha_i + \beta_j + \gamma_{ij}$$

are not uniquely determined, because we can add and subtract constants on the right-hand side, changing the parameters but maintaining equality. It is necessary to impose constraints for uniqueness, namely,

$$\sum_{i=1}^{I} \alpha_i = 0, \quad \sum_{j=1}^{J} \beta_j = 0, \quad \sum_{i=1}^{I} \gamma_{ij} = 0, \quad \sum_{j=1}^{I} \gamma_{ij} = 0$$

We can then **interpret** α_i as the contribution or deviation of level i of factor A from a baseline (of 0 in view of $\sum_{i=1}^{I} \alpha_i = 0$). Likewise, β_j is the contribution of level j of factor B and γ_{ij} is any possible interaction of combination (i, j).

If we fix attention on level i of factor A and average over all levels of factor B, we can estimate $\mu + \alpha_i$ using the sample mean of all observations with the same value i,

$$\frac{1}{Jn} \sum_{j=1}^{J} \sum_{k=1}^{n} x_{ijk} = \bar{x}_{i\bullet}$$

We can also estimate $\mu + \beta_j$ using

$$\frac{1}{In} \sum_{i=1}^{I} \sum_{k=1}^{n} x_{ijk} = \bar{x}_{\bullet j}$$

We then take differences and find the natural estimates (recalling that $\bar{x}$ is the overall mean):

$$
\begin{aligned}
\hat{\alpha}_i &\equiv \bar{x}_{i\bullet} - \bar{x} & \text{estimates } \alpha_i \\
\hat{\beta}_i &\equiv \bar{x}_{\bullet j} - \bar{x} & \text{estimates } \beta_j \\
\hat{\gamma}_{ij} &\equiv \bar{x}_{ij} - \bar{x}_{i\bullet} - \bar{x}_{\bullet j} + \bar{x} & \text{estimates } \gamma_{ij}
\end{aligned}
$$

Then a minor miracle occurs. Each of these estimates contributes to a marvelous sum of squares decomposition; namely, if we define

$$
\text{SSA} = \sum_{i=1}^{I} \sum_{j=1}^{J} \sum_{k=1}^{n} \hat{\alpha}_i^2
$$

$$
\text{SSB} = \sum_{i=1}^{I} \sum_{j=1}^{J} \sum_{k=1}^{n} \hat{\beta}_j^2
$$

$$
\text{SSAB} = \sum_{i=1}^{I} \sum_{j=1}^{J} \sum_{k=1}^{n} \hat{\gamma}_{ij}^2
$$

then

$$
\text{SSM} = \text{SSA} + \text{SSB} + \text{SSAB}
$$

Each term on the right carries a degrees of freedom

$$
\begin{aligned}
\text{DFA} &= I - 1 \\
\text{DFB} &= J - 1 \\
\text{DFAB} &= (I-1)(J-1)
\end{aligned}
$$

giving corresponding mean squares and F ratios as in one-way ANOVA. For instance,

$$
\begin{aligned}
\text{MSA} &= \frac{\text{SSA}}{I-1} \\
F &= \frac{\text{MSA}}{\text{MSE}}
\end{aligned}
$$

and this particular F ratio is used to test

$$
\begin{aligned}
H_{0A} &: \alpha_i = 0, \quad 1 \le i \le I \\
H_{aA} &: \text{at least two } \alpha_i \text{ are not zero}
\end{aligned}
$$

The decision rule is that if

$$
F = \frac{\text{MSA}}{\text{MSE}} \quad \text{exceeds the critical value } F^*(I-1, N-IJ)
$$

then we reject H_{0A} and conclude that there is a difference among the means for the levels of factor A; that is, factor A is significant. An analogous analysis is carried out for factor B and then for interaction if neither factor A nor factor B is judged significant.

13.2   Inference for Two-Way ANOVA

While the theory just described may seem intimidating, the practical implementation of the procedure could not be simpler once the data are properly recorded in the workbook. Excel outputs summary statistics, sums of squares, mean squares, F ratios, critical F values, and P-values in an extensive ANOVA table.

Table 13.1: Iron Content

Type of Pot	Type of Food		
	Meat	Legumes	Meat
Aluminum	1.77	2.40	1.03
	2.36	2.17	1.53
	1.96	2.41	1.07
	2.14	2.34	1.30
Clay	2.27	2.41	1.55
	1.28	2.43	0.79
	2.48	2.57	1.68
	2.68	2.48	1.82
Iron	5.27	3.69	2.45
	5.17	3.43	2.99
	4.06	3.84	2.80
	4.22	3.72	2.92

Example 13.1.   (Exercise 13.16, page 822 in the text.) Iron deficiency anemia is the most common form of malnutrition in developing countries, affecting about 50% of children and women and 25% of men. Iron pots for cooking foods had traditionally been used in many of these countries, but they have been largely replaced by aluminum pots, which are cheaper and lighter. Some research has suggested that food cooked in iron pots will contain more iron than food cooked in other types of pots. One study designed to investigate this issue compared the iron content of some Ethiopian foods cooked in aluminum, clay, and iron pots. The iron content of *yesiga wet'*, beef cut into small pieces and prepared with several Ethiopian spices, *shiro wet'*, a legume-based mixture of chickpea flour and Ethiopian spiced pepper, and *ye-atkilt allych'a*, a lightly spiced vegetable casserole was measured. Four samples of each food were cooked in each type of pot. Table 13.1 gives the iron content in milligrams of iron per 100 grams of cooked food.

(a) Make a table giving the sample size, mean, and standard deviation for each type of pot.

(b) Plot the means and give a short summary of how the iron content of foods depends upon the cooking pot.

(c) Run the analysis of variance. Give the ANOVA table, the F- statistics with degrees of freedom and P-values, and your conclusions regarding the hypotheses about main effects and interactions.

The Two-Way ANOVA Tool

The explanatory variable of interest is the type of pot while another variable, type of food, has been used as a blocking variable. This is equivalent to a two-way anova and we will therefore refer to "Pot" as Factor A (3 levels: Aluminum, Clay, Iron) and "Food" as Factor B (3 levels: Meat, Legumes, Vegetables). There are 4 replications per treatment so $n = 4$ for a total of $N = I \times J \times n = 3 \times 3 \times 4 = 36$ observations.

	A	B	C	D
1	Two-Way Analysis of Variance			
2				
3	Pot		Food	
4		Meat	Legumes	Vegetables
5	Aluminum	1.77	2.40	1.03
6		2.36	2.17	1.53
7		1.96	2.41	1.07
8		2.14	2.34	1.30
9	Clay	2.27	2.41	1.55
10		1.28	2.43	0.79
11		2.48	2.57	1.68
12		2.68	2.48	1.82
13	Iron	5.27	3.69	2.45
14		5.17	3.43	2.99
15		4.06	3.84	2.80
16		4.22	3.72	2.92

Figure 13.1: Preparing the Data

Preparing the Data

1. Enter the data exactly as shown in Fig 13.1. Excel will balk if the data are not laid out correctly.

2. From the Menu Bar, choose **Tools – Data Analysis – Anova: Two-Factor With Replication** and complete the dialog box with entries as shown in Fig 13.2. The output appears in cells E1:K36.

Excel Output—Two-Way ANOVA

There are two components to the output in Fig 13.3. The top portion provides summary statistics such as sample means and variances, which are useful in plotting. The lower half is the ANOVA table.

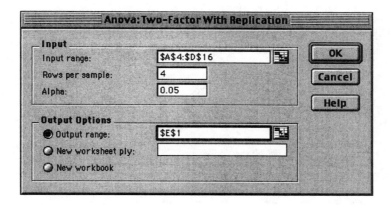

Figure 13.2: Two-Way ANOVA Dialog Box

Summary

Referring to Fig 13.3, the block E3:H20 gives summary statistics for each treatment labelled by a level of Factor A in column E and a level of Factor B in row 3. For instance for the treatment (Aluminum × Meat) we read in cell F5 there are 4 observations, from cell F7 the average $x_{11} = 2.0575$, and from cell F8 the variance $s_{11}^2 = 0.063492$, and so on. Standard deviations are not given but may be calculated from the variances.

The column block on the right labeled "Total" (I3:G20) provides summaries for each level of factor A (summed or averaged over all levels of factor B). For instance the average iron content for iron pots is shown in cell I19 as $\bar{x}_{3\bullet} = 3.7133$. This is the average of the 12 observations corresponding to iron pots. Likewise the average iron content for Aluminum pots is $\bar{x}_{1\bullet} = 1.8733$ in cell I7, and for Clay pots the average iron content is $\bar{x}_{2\bullet} = 2.0367$ in cell I13.

The corresponding summaries for factor B are in the row block headed by the label Total in E22:E26. For example, the average for the 12 observations for "Meat" is $\bar{x}_{\bullet 1} = 2.9717$ in cell F25.

ANOVA Table

There are three separate possible significance tests: Factor A, Factor B, and Interaction. Block E29:K36 presents the ANOVA table listing the four sources of variation (with the terminology: Sample ≡ Factor A; Column ≡ Factor B; Interaction, Within ≡ Error; and Total), their corresponding sums of squares, degrees of freedom, mean square, computed F, P-value, and critical F^* values.

We can immediately read off the conclusions. First, there does not appear to be any significant interaction effect, so we may then examine the two factors individually. For Factor A, since the corresponding computed F statistic 92.263 exceeds the critical $F^*(2, 27) = 3.354$ we conclude that there is a difference in iron

	E	F	G	H	I	J	K
1	Anova: Two-Factor With Replication						
2							
3	SUMMARY	Meat	Legumes	Vegetables	Total		
4	*Aluminum*						
5	Count	4	4	4	12		
6	Sum	8.23	9.32	4.93	22.48		
7	Average	2.0575	2.3300	1.2325	1.8733		
8	Variance	0.063492	0.012333	0.053492	0.272770		
9							
10	*Clay*						
11	Count	4	4	4	12		
12	Sum	8.71	9.89	5.84	24.44		
13	Average	2.1775	2.4725	1.4600	2.0367		
14	Variance	0.386025	0.005092	0.211667	0.361606		
15							
16	*Iron*						
17	Count	4	4	4	12		
18	Sum	18.72	14.68	11.16	44.56		
19	Average	4.6800	3.6700	2.7900	3.7133		
20	Variance	0.394733	0.029800	0.057533	0.781970		
21							
22	*Total*						
23	Count	12	12	12			
24	Sum	35.66	33.89	21.93			
25	Average	2.9717	2.8242	1.8275			
26	Variance	1.824724	0.406808	0.602730			
27							
28							
29	ANOVA						
30	*Source of Variation*	SS	df	MS	F	P-value	F crit
31	Sample	24.89396	2	12.44698	92.263	0.00000	3.354
32	Columns	9.29687	2	4.64844	34.456	0.00000	3.354
33	Interaction	2.64043	4	0.66011	4.893	0.00425	2.728
34	Within	3.6425	27	0.13491			
35							
36	Total	40.47376	35				

Figure 13.3: Two-Way ANOVA Output

content among the three types of pots. We also conclude that there is a difference in iron content due to Factor B, the type of food. This was to be expected and shows that blocking on this factor was effective.

Profile (Interaction) Plot

A profile plot is a simple graphical diagnostic tool for displaying the numerical summaries in the Excel output. It is handy for seeing possible interactions visually. A profile plot is a graph of all the treatment means in a manner which gives some visual insight. The sample means of all the treatments are plotted on the y-axis against the corresponding levels of one of the factors on the x-axis. The following steps show how to use the **ChartWizard** and the summary output to produce a profile plot. While it really does not matter which factor is used for x since the same information is displayed in either case, still some features may be more apparent visually in one graph than in the other. Since "Pot" is the primary factor of interest we will plot the sample means against the types of pots on the x axis.

	N	O	P	Q
1		**Profile Plot of Means**		
2				
3	Pot		Food	
4		Meat	Legumes	Vegetables
5	Aluminum	2.058	2.330	1.233
6	Clay	2.178	2.473	1.460
7	Iron	4.680	3.670	2.790

Figure 13.4: Interactions—Data

1. Enter the data as in the top half of Fig 13.4.

2. **Users of Excel 5/95**

   - Click the **ChartWizard** button. In Step 1 enter the range N4:Q7, in Step 2 select **Line** chart, in Step 3 select Format **1**, in Step 4 select Data Series in **Rows**, Row "1" for Category(X) Labels, Use Column "1" for Legend Text, and finally in Step 5 select **Yes** for Add a Legend? and type the name for the chart and the axis titles.

   Users of Excel 97/98/2000/2001

   - Select the range N4:Q7 and click the **ChartWizard** button. In Step 1 select **Line** for Chart Type and the upper left Chart sub-type **Line**. In Step 2 under the **Data Range** tab, enter K3:O6 and select the Series radio button for **Rows** (Note what happens in the sample display if you select the radio button **Columns**. In Step 3 enter the titles under the **Titles** tab, remove gridlines under the **Gridlines** tab, check **Show legend** under the **Legend** tab, and under the **Data Labels** tab select the radio button **None**. In Step 4 click Finish.

3. Finally, make any additional editing changes so that the resulting chart appears similar to Fig 13.5.

Output

The profile plot shows that the sample means for iron pots are all larger than the sample means for the other types of pots. Also notice that there is a crossover in the lines. Legumes which show a consistently higher iron content than meat or vegetables in aluminum and clay pots are second highest in iron content in the iron pots. Such a pattern indicates the possibility that interaction is present. However the F test for ANOVA did not indicate this interaction. The apparent contradiction results because the estimated standard deviation (s_p^2 from cell H34) $s_p = \sqrt{0.13491} = 0.367$ is so large that such a crossover in means can be explained as random variation. This example indicates that some caution needs to be exercised in interpreting profile plots.

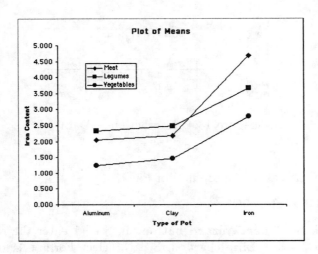

Figure 13.5: Profile Plot vs. Pots

Exercise. By selecting the Series radio button for **Rows** in the **ChartWizard** produce the profile plot shown in Fig 13.6 with type of food on the horizontal axis. The same nine sample means are plotted, but with different abscissae than in Fig 13.5.

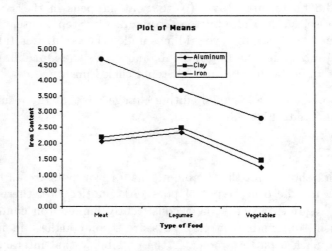

Figure 13.6: Profile Plot vs. Food

Chapter 14

Nonparametric Tests

Nonparametric procedures are based on the ranks of observations and replace assumptions of normality with less stringent assumptions, such as symmetry and continuity of distributions. Excel does not provide nonparametric tests. Nonetheless, these tests may readily be developed.

14.1 The Wilcoxon Rank Sum Test

The Wilcoxon rank sum test is a procedure for comparing independent samples from two populations. Suppose

$$(x_{11}, x_{12}, \ldots, x_{1n_1}), (x_{21}, x_{22}, \ldots, x_{2n_2})$$

are the two samples. Generally sample 1, the x_{1j} observations, might be the control group, while sample 2, the x_{2j} observations, might be the treatment group.

We wish to test whether both samples can be assumed to arise from identical populations or whether the populations are shifted by a constant. The precise assumption is that if F_i represents the cumulative distribution function of the x_i-observations, $1 \leq i \leq 2$, then

$$F_2(t) = F_1(t + \Delta) \qquad \text{for all real } t$$

where Δ is a constant representing an unknown shift in the two distributions. If Δ is positive, then the second sample would contain systematically higher values than the first sample.

The null hypothesis is thus

$$H_0 : \mu_1 - \mu_2 = 0$$

where μ_1 and μ_2 are the two population means (or medians).

It is also possible to test a nonzero difference

$$H_0 : \mu_1 - \mu_2 = \Delta_0$$

by subtracting Δ_0 from each sample 1 observation and applying the first test for a zero difference in means to

$$(x_{11} - \Delta_0, x_{12} - \Delta_0, \ldots, x_{1n_1} - \Delta_0), (x_{21}, x_{22}, \ldots, x_{2n_2})$$

Description of the Procedure

1. Rank all the observations from smallest to largest into one list with $N = n_1 + n_2$ observations.

2. Let r_j be the rank of the observation x_{1j}.

3. Set

$$W = \sum_{j=1}^{n_1} r_j$$

which is the sum of the ranks associated with the data from sample 1.

If H_0 is true, then each overall ranking of the N combined observations would have the same probability, but if, for instance,

$$H_a : \mu_1 - \mu_2 > 0$$

then the ranks of sample 1, contributing to W and arising from the population with the larger mean under H_a, would be larger than expected under H_0, leading to an observed W value above the mean. The Wilcoxon rank sum test therefore rejects H_0 if W is beyond some reasonable value. This test is sometimes called the Mann-Whitney test because it was originally derived as an alternative, but equivalent formulation by H. B. Mann and D. R. Whitney in 1947.

The exact distribution of W has been tabulated, but we will base our procedure on the normal approximation. It can be shown that if H_0 is true, then

$$\mu_W \;=\; \text{mean of } W = \frac{n_1(n_1 + n_2 + 1)}{2}$$

$$\sigma_w \;=\; \text{standard deviation of } W = \sqrt{\frac{n_1 n_2(n_1 + n_2 + 1)}{12}}$$

Calculate

$$z = \frac{W - \mu_W}{\sigma_W}$$

For a fixed level α test:

$$\begin{array}{ll}
\text{if } H_a : \mu_1 - \mu_2 > 0 & \text{reject } H_0 \text{ if } z > z^* \\
\text{if } H_a : \mu_1 - \mu_2 < 0 & \text{reject } H_0 \text{ if } z < -z^* \\
\text{if } H_a : \mu_1 - \mu_2 \neq 0 & \text{reject } H_0 \text{ if } |z| > z^*
\end{array}$$

where z^* represents the corresponding upper critical value of a standard normal distribution.

The Wilcoxon Rank Sum Test in Practice

Example 14.1. (Examples 14.1–14.5, pages 3-10 in the additional chapters in the Student CD-ROM.) Does the presence of small numbers of weeds reduce the yield of corn? Lamb's-quarter is a common weed in corn fields. A researcher planted corn at the same rate in eight small plots of ground, then weeded the corn rows by hand to allow no weeds in four randomly selected plots and exactly three lamb's-quarter plants per meter of row in the other four plots. Here are the yields of corn (bushels per acre) in each of the plots.

Weeds per meter	Yield (bu/acre)			
0	166.7	172.2	165.0	176.9
3	158.6	176.4	153.1	156.0

Normal quantile plots suggest that the data may be right-skewed. The samples are too small to assess normality adequately or to rely on the robustness of the two-sample t test. We may prefer to use a test that does not require normality. Carry out the significance test

H_0 : no difference in distribution of yields

H_a : yields are systematically higher in weed-free plots

Solution. Fig 14.1 shows the Excel formulas required and the corresponding values taken when applied to the above data set.

1. Enter the labels "Sample 1," "Sample 2," "Population," "Combined," and "Rank" in cells A3:E3.

2. Record the observations in Sample 1 in cells A4:A7 and copy them to D4:D7. Record the observations in Sample 2 in cells B4:B7 and copy them to D8:D11. Enter the value 1 in C4:C7, the value 2 in C8:C11, and finally the values $\{1, 2, 3, 4, 5, 6, 7, 8\}$ in E4:E11.

3. **Name** the ranges for Sample 1, Sample 2, Population, Combined, and Rank.

4. **Rank** the combined data set that has been copied into D4:D11, and carry the corresponding sample numbers (Sample 1 = 0 weeds per meter, Sample 2 = 3 weeds per meter) back to C4:C11 as follows. Select C3:D11 and from the Menu Bar choose **Data – Sort** to bring up the **Sort** dialog box. Select the radio buttons for "Ascending" and for "Header row." Then click OK. This ranks the combined sample in D4:D11 in increasing order and also carries the corresponding sample labels in C4:C11. Fig 14.1 shows the results of the sort.

	A	B	C	D	E	F
1				**Wilcoxon Rank Sum Test**		
2						
3	Sample1	Sample2	Population	Combined	Rank	
4	166.7	158.6	2	153.1	1	
5	172.2	176.4	2	156.0	2	
6	165.0	153.1	2	158.6	3	
7	176.9	156.0	1	165.0	4	
8			1	166.7	5	
9			1	172.2	6	
10			2	176.4	7	
11			1	176.9	8	
12						
13	User Input					
14	alpha	0.05				
15	Summary Statistics					
16	n_1	4		=COUNT(Sample1)		
17	n_2	4		=COUNT(Sample2)		
18	Calculations					
19	W	23		=SUMIF(Population, "=1", Rank)		
20	mu	18		=n_1*(n_1+n_2+1)/2		
21	sigma	3.464		=SQRT(n_1*n_2*(n_1+n_2+1)/12)		
22	z	1.443		=(W-mu)/sigma		
23	Lower Test					
24	lower_z			=NORMSINV(alpha)		
25	Decision			=IF(z<lower_z,"Reject H0","Do Not Reject H0")		
26	Pvalue			=NORMSDIST(z)		
27	Upper Test					
28	upper_z	1.645		=-NORMSINV(alpha)		
29	Decision	Do Not Reject H0		=IF(z>upper_z,"Reject H0","Do Not Reject H0")		
30	Pvalue	0.0745		=1-NORMSDIST(z)		
31	Two-Sided Test					
32	two_z			=ABS(NORMSINV(alpha/2))		
33	Decision			=IF(ABS(z)>two_z,"Reject H0","Do Not Reject H0")		
34	Pvalue			=2*(1-NORMSDIST(ABS(z)))		

Figure 14.1: Wilcoxon Rank Sum Test—Formulas and Values

5. Next we carry out calculations akin to those in Chapter 6 (see Fig 6.2) for a one-sample test of a normal mean. Enter the labels as shown in cells A13:A34 on Fig 14.1, and **Name** the corresponding ranges in the respective column B cells to be able to refer to n_1, n_2, W, mu, sigma, z, lower_z, upper_z, and two_z by name in the ensuing formulas shown. The formulas to be entered in column B are presented in column D, and the values taken when the formulas are applied to this data (that is, what you will see in your workbook) are shown in the corresponding cells in column B. For instance, enter the formula from cell D19

$$= \texttt{SUMIF}(\text{Population}, \text{``}= 1\text{''}, \text{Rank})$$

into cell B19. This Excel function adds those cells under Rank whose corresponding Population is 1. In other words, the function adds the ranks corresponding to observations from Sample 1. In cell B19 appears the answer 23, the value of the Wilcoxon rank sum statistic. Formulas for all three types of alternate hypotheses are provided in Fig 14.1, but the values are shown only for the particular alternative in this problem (upper test).

When using this template, remember to use only the appropriate cells for the problem at hand.

Interpreting the Results

We read off

$$
\begin{aligned}
W &= 23 \quad &\text{(cell B19)} \\
\mu_W &= 18 \quad &\text{(cell B20)} \\
\sigma_W &= 3.464 \quad &\text{(cell B21)} \\
z \text{ statistic} &= 1.443 \quad &\text{(cell B22)}
\end{aligned}
$$

The alternate hypothesis is

$$H_a : \mu_1 - \mu_2 > 0$$

so that rows 27–30 are appropriate (rows 23–26 for a lower-tailed test and rows 31–34 for a two-tailed test). We find that

$$
\begin{aligned}
\text{upper critical value } z^* &= 1.645 \quad &\text{(cell B28)} \\
\text{decision rule} &= \text{``Do not reject''} \quad &\text{(cell B29)} \\
P\text{-value} &= 0.0745 \quad &\text{(cell B30)}
\end{aligned}
$$

We conclude that the data are not significant at the nominal 5% level of significance.

Continuity Correction for the Normal Approximation

A more accurate P-value is obtained by applying the continuity correction, which adjusts for the fact that a continuous distribution (the normal) is being used to approximate a discrete distribution W. If we use a correction of 0.5, then the upper-tailed test statistic becomes

$$z = \frac{W - 0.5 - \mu_W}{\sigma_W}$$

and this leads to

$$z = 1.299$$

and

$$P\text{-value} = 1 - \Phi(1.299) = 0.0970$$

Note: These more accurate values appear in Examples 14.4 and 14.5 on page 2 in the Student CD-ROM.

Ties

Theoretically, the assumption of a continuous distribution ensures that all $n_1 + n_2$ observed values will be different. In practice, ties are sometimes observed. The common practice is to average the ranks for the tied observations and carry on as above with a change in the standard deviation. Use

$$\sigma_W^2 = \frac{n_1 n_2}{12} \left(n_1 + n_2 + 1 - \frac{\sum_{i=1}^{G} t_i(t_i^2 - 1)}{(n_1 + n_2)(n_1 + n_2 - 1)} \right)$$

where G is the number of tied groups and t_i is the number of tied observations in the ith tied group. Unless G is large, the adjustment in the formula for the variance makes little difference.

14.2   The Wilcoxon Signed Rank Test

The Wilcoxon signed rank test is a nonparametric version of the one-sample procedures based on the assumption of normal population discussed in Chapters 6 and 7. The key assumption is that the data arise from a population symmetric about its mean.

For this reason one of its most useful applications is in a matched-pairs setting with n pairs (x_{1i}, x_{2i}) of observations, where it is natural to assume that the populations from which the pairs are taken differ only by a shift in the mean (that is, the population distribution shapes are otherwise the same). The differences then satisfy the requirement of symmetry under the null hypothesis of equality of means.

Description of the Matched-Pairs Procedure

The data consist of n pairs (x_{1i}, x_{2i}) of observations. The $\{x_{1i}\}$ are a sample from a population with mean μ_1 and the $\{x_{2i}\}$ are a sample from a population with mean μ_2. The null hypothesis is

$$H_0 : \mu_1 - \mu_2 = 0$$

1. Form the absolute differences $|d_j|$, where $d_j = x_{1j} - x_{2j}$.

2. Let r_j be the rank of $|d_j|$ in the joint ranking of the $\{|d_j|\}$, from smallest to largest.

3. Form the sum of the positive signed ranks

$$W^+ = \sum r_j$$

where the sum is taken over all ranks r_j for which the corresponding difference d_j is positive.

The Wilcoxon signed rank procedure rejects H_0 if W^+ is beyond some reasonable value, in particular for values of W^+ that are too large or too small.

As with the Wilcoxon rank sum statistic W, there exist tables of the exact distribution of W^+, but we will base our procedure on the normal approximation. When H_0 is true the mean and the standard deviation of W^+ are given by

$$\mu_{W^+} = \frac{n(n+1)}{4}$$

$$\sigma_{W^+} = \sqrt{\frac{n(n+1)(2n+1)}{24}}$$

We then calculate

$$z = \frac{W^+ - \mu_{W^+}}{\sigma_{W^+}}$$

and for a fixed level α test:

$$\text{if } H_a : \mu_1 - \mu_2 > 0 \quad \text{reject } H_0 \text{ if } z > z^*$$
$$\text{if } H_a : \mu_1 - \mu_2 < 0 \quad \text{reject } H_0 \text{ if } z < -z^*$$
$$\text{if } H_a : \mu_1 - \mu_2 \neq 0 \quad \text{reject } H_0 \text{ if } |z| > z^*$$

The Wilcoxon Signed Rank Test in Practice

Example 14.2. (Examples 14.8–14.10, pages 18–21 in the Student CD-ROM.) A study of early childhood education asked kindergarten students to tell two fairy tales that had been read to them earlier in the week. The first tale had been read to them and the second had been read but also illustrated with pictures. An expert listened to a recording of the children and assigned a score for certain uses of language. Here are the data for five "low progress" readers in a pilot study:

Child	1	2	3	4	5
Story 2	0.77	0.49	0.66	0.28	0.38
Story 1	0.40	0.72	0.00	0.36	0.55
Difference	0.37	−0.23	0.66	−0.08	−0.17

We wonder if illustrations improve how the children retell a story. We would like to test the hypotheses

H_0 : scores have the same distribution for both stories

H_a : scores are systematically higher for story 2

Because this is a matched-pairs design, we base our inference on the differences. The matched-pairs t test gives $t = 0.635$ with a one-sided P-value of 0.280. As displays of the data suggest a mild lack of normality, carry out a nonparametric significance test.

Solution.   Fig 14.2 shows the Excel formulas required and the corresponding values taken for the example data set. The calculations are similar to those in the previous section.

	A	B	C	D	E	F	G
1			Wilcoxon Signed Rank Test – Matched Pairs				
2							
3	Sample1	Sample2	Diff		Ranked_Diff	Ranked_Abs_Diff	Rank
4	0.77	0.40	0.37		-0.08	0.08	1
5	0.49	0.72	-0.23		-0.17	0.17	2
6	0.66	0.00	0.66		-0.23	0.23	3
7	0.28	0.36	-0.08		0.37	0.37	4
8	0.38	0.55	-0.17		0.66	0.66	5
9							
10							
11	User Input						
12	alpha	0.05					
13	Summary Statistics						
14	n	5		=COUNT(Sample1)			
15	Calculations						
16	W+	9		=SUMIF(Ranked_Diff, ">0", Rank)			
17	mu	7.5		=n*(n+1)/4			
18	sigma	3.708		=SQRT(n*(n+1)*(2*n+1)/24)			
19	z	0.405		=(Wplus-mu)/sigma			
20	Lower Test						
21	lower_z			=NORMSINV(alpha)			
22	Decision			=IF(z<lower_z,"Reject HO","Do Not Reject HO")			
23	Pvalue			=NORMSDIST(z)			
24	Upper Test						
25	upper_z	1.645		=-NORMSINV(alpha)			
26	Decision	Do Not Reject HO		=IF(z>upper_z,"Reject HO","Do Not Reject HO")			
27	Pvalue	0.3429		=1-NORMSDIST(z)			
28	Two-Sided Test						
29	two_z			=ABS(NORMSINV(alpha/2))			
30	Decision			=IF(ABS(z)>two_z,"Reject HO","Do Not Reject HO")			
31	Pvalue			=2*(1-NORMSDIST(ABS(z)))			

Figure 14.2: Wilcoxon Signed Rank—Matched Pairs

1. Enter the labels "Sample 1," "Sample 2," and "Diff" in cells A3:C3 and the labels "Ranked_Diff," "Ranked_Abs_Diff," and "Rank" in cells E3:G3.

2. Record the observations in Sample 1 in cells A4:A8 and the observations for Sample 2 in cells B4:B8. In cells C4:C8 record the difference between Sample 1 and Sample 2. Enter the values $\{1, 2, 3, 4, 5\}$ in G4:G8.

3. **Name** the ranges for the corresponding labels Sample 1, Sample 2, Ranked_Diff, Ranked_Abs_Diff, and Rank to include the respective cells in rows 4–8.

4. In a different part of the sheet, copy the values of the differences (not their formulas, if Excel calculated them) and then enter their absolute values using the Excel function **ABS()**, which gives the absolute value of its argument. As in step 4 of the previous section, **rank** the absolute values of the differences in increasing order and carry along the actual differences. Copy the results to cells E4:F8 as shown in Fig 14.2, showing the ranked absolute differences in

column F and the actual ranked differences in column E. We need the latter to recognize which absolute differences correspond to positive differences in calculating W^+.

5. The calculations required are shown in the lower portion of Fig 14.2, where we have included in column D the formulas that are to be entered in column B. The actual values taken by these formulas—the values that will appear on your workbook—are in column B. Refer to step 5 of the previous section for the analogous details, and remember to name all ranges used in the formulas shown.

Interpreting the Results

We read off

$$
\begin{aligned}
W^+ &= 9 &\text{(cell B16)} \\
\mu_{W+} &= 7.5 &\text{(cell B17)} \\
\sigma_{W+} &= 3.708 &\text{(cell B18)} \\
z\text{statistic} &= 0.405 &\text{(cell B19)}
\end{aligned}
$$

The alternate hypothesis requires an upper-tailed test for which

$$
P\text{-value} = 0.3429 \qquad \text{(cell B27)}
$$

We conclude that the data are not significant.

Continuity Correction for the Normal Approximation

As with the Wilcoxon rank sum test, a more accurate P-value is obtained with the continuity correction

$$
z = \frac{W^+ - 0.5 - \mu_{W+}}{\sigma_{W+}}
$$

and this leads to

$$
z = 0.270
$$

$$
P\text{-value} = 1 - \Phi(0.270) = 0.3937
$$

Note: These more accurate values are shown in Example 14.10 on page 21 in the Student CD-ROM.

Ties and Zero Values

If there are zeros among the differences $\{d_i\}$, discard them and use for n the number of nonzero $\{d_i\}$. If there are any ties, then use the average rank for each set of tied observations and apply the procedure with variance

$$\sigma_{W+}^2 = \frac{1}{24}\left(n(n+1)(2n+1) - \frac{\sum_{i=1}^{G} t_i(t_i^2 - 1)}{2}\right)$$

where G is the number of tied groups and t_i are the number of tied observations in the ith tied group.

14.3   The Kruskal-Wallis Test

In this section we generalize the Wilcoxon rank sum test to situations involving independent samples from I populations when the assumptions required for validity of the one-way ANOVA in Chapter 12 cannot be substantiated.

The data consist of $N = \sum_{i=1}^{I} n_i$ observations with $n_i \geq 1$ observations $\{x_{ij} : 1 \leq j \leq n_i\}$ taken from population i. The assumption replacing normality is

$$x_{ij} = \mu_i + \varepsilon_{ij} \quad 1 \leq i \leq I, \ 1 \leq j \leq n_i$$

where the errors $\{\varepsilon_{ij}\}$ are mutually independent with mean 0 and have the same *continuous* distribution. If we let $F(x)$ be the cumulative distribution function (c.d.f.) of a generic error term, this assumption is tantamount to $F_i(x)$ being the c.d.f. of population i, where

$$F_i(x) \equiv F(x - \mu_i) \quad 1 \leq i \leq I$$

The significance test is

$$H_0 : \mu_1 = \mu_2 = \cdots = \mu_I$$
$$H_a : \text{not all of the } \mu_i \text{ are equal}$$

and the procedure generalizing the rank sum test is called the Kruskal-Wallis test.

Description of the Procedure

1. **Rank** all the observations jointly from smallest to largest.

2. Let r_{ij} be the rank of observation x_{ij}.

3. Set

$$R_i = \sum_{j=1}^{n_i} r_{ij}$$

which is the sum of the ranks associated with sample i.

Denote by

$$\bar{R}_i = \frac{1}{n_i} R_i$$

the average rank in sample i. If H_0 is true, then by symmetry the mean of any rank r_{ij} is $E(r_{ij}) = \frac{N+1}{2}$, which is the average of the integers $\{1, 2, \ldots, N\}$, and therefore $E[\bar{R}_i] = E\left[\frac{1}{n_i} R_i\right] = E\left[\frac{1}{n_i} \sum_{j=1}^{n_i} r_{ij}\right] = \frac{N+1}{2}$. Thus, we would expect the ranks to be uniformly intermingled among the I samples. But if H_0 is false, then some samples will tend to have many small ranks, while others will have many large ranks. Just as in ANOVA, we take the sum of squares of the differences between the average rank $\bar{R}_i$ of each sample and the overall average $\frac{N+1}{2}$ by computing

$$\frac{12}{N(N+1)} \sum_{i=1}^{I} n_i \left(\bar{R}_i - \frac{N+1}{2}\right)^2$$

which can be expressed equivalently as

$$H = \frac{12}{N(N+1)} \sum_{i=1}^{I} \frac{R_i^2}{n_i} - 3(N+1)$$

and called the Kruskal-Wallis statistic. We then reject H_0 for "large" values of H.

Tables of critical values exist for small values of the $\{n_i\}$, but it is customary to use a normal approximation, which provides an approximate sampling distribution:

H is approximately chi-square with $I - 1$ degrees of freedom.

Therefore the test is

Reject H_0 if $H > \chi^2$

where χ^2 is the upper critical α value of a chi-square distribution on $I - 1$ degrees of freedom.

The Kruskal-Wallis Test in Practice

Example 14.3. (Examples 14.13 and 14.14, pages 26–30 in the Student CD-ROM.) Lamb's-quarter is a common weed that interferes with the growth of corn. A researcher planted corn at the same rate in 16 small plots of ground, then randomly assigned the plots to four groups. He weeded the plots by hand to allow a fixed number of lamb's-quarter plants to grow in each meter of corn row. These numbers were 0, 1, 3, and 9 in the four groups of plots. No other weeds were allowed to grow, and all plots received identical treatment, except for the weeds. Here are the yields of corn (bushels per acre) in each of the plots.

Weeds per meter	Corn yield	Weeds per meter	Corn yield	Weeds per meter	Corn yield	Weeds per meter	Corn yield
0	166.7	1	166.2	3	158.6	9	162.8
0	172.2	1	157.3	3	176.4	9	142.4
0	165.0	1	166.7	3	153.1	9	162.7
0	176.9	1	161.1	3	156.0	9	162.4

The summary statistics follow:

Weeds	n	Mean	Std Dev
0	4	170.200	5.422
1	4	162.825	4.469
3	4	161.025	10.498
9	4	157.575	10.118

The sample standard deviations do not satisfy our rule of thumb that for safe use of ANOVA the largest should not exceed twice the smallest. Normal quantile plots show that outliers are present in the yields for 3 and 9 weeds per meter. Use the Kruskal-Wallis procedure to test

H_0 : yields have the same distribution in all groups

H_a : yields are systematically higher in some groups than others.

Solution.   Fig 14.3 shows the Excel formulas required and the corresponding values taken when applied to the example data set.

As we have already described in detail the construction of the two earlier nonparametric procedures, we leave as an exercise the application of this workbook. Beginning in row 9, column C shows the formulas to be entered into the adjacent cells of column B, where the numerical evaluation of the formulas is shown.

	A	B	C	D	E	F	G	H
1				**Kruskal-Wallis Test**				
2								
3	Sample1	Sample2	Sample3	Sample4		Pop	Combined	Rank
4	166.7	166.2	158.6	162.8		4	142.4	1
5	172.2	157.3	176.4	142.4		3	153.1	2
6	165.0	166.7	153.1	162.7		3	156.0	3
7	176.9	161.1	156.0	162.4		2	157.3	4
8						3	158.6	5
9	R_1	52	=SUMIF(Pop, "=1", Rank)			2	161.1	6
10	R_2	34	=SUMIF(Pop, "=2", Rank)			4	162.4	7
11	R_3	25	=SUMIF(Pop, "=3", Rank)			4	162.7	8
12	R_4	25	=SUMIF(Pop, "=4", Rank)			4	162.8	9
13						1	165.0	10
14	n_1	4	=COUNT(A4:A7)			2	166.2	11
15	n_2	4	=COUNT(B4:B7)			1	166.7	12
16	n_3	4	=COUNT(C4:C7)			2	166.7	13
17	n_4	4	=COUNT(D4:D7)			1	172.2	14
18	N	16	=n_1+n_2+n_3+n_4			3	176.4	15
19						1	176.9	16
20		676	=R_1^2/n_1					
21		289	=R_2^2/n_2					
22		156.25	=R_3^2/n_3					
23		156.25	=R_4^2/n_4					
24								
25	H	5.3602941	=(12/(N*(N+1)))*SUM(B21:B24) – 3*(N+1)					
26	Critical 5%	7.815	=CHIINV(0.05,3)					
27	P-value:	1.47E-01	=CHIDIST(H,3)					

Figure 14.3: Kruskal-Wallis Test—Formulas and Values

Interpreting the Results

We read off

$$
\begin{aligned}
H &= 5.36 \quad \text{(cell B25)} \\
\chi^2 &= 7.815 \quad \text{(cell B26)} \\
P\text{-value} &= 0.147 \quad \text{(cell B27)}
\end{aligned}
$$

The data are not significant at the 5% level, meaning that there is no convincing evidence that more weeds decrease yield.

Ties

We have ignored a tie in the above calculation; the value 166.7 appears in Sample 1 and in Sample 2. For a more accurate calculation, we give to the value 166.7 the average rank 12.5. Then replace H with

$$
H' = \frac{H}{1 - \sum_{i=1}^{G} \frac{t_i(t_i^3 - 1)}{N^3 - N}}
$$

where G is the number of tied groups and t_i are the number of tied observations in the ith tied group.

Exercise. Replace the ranks of the two observations 166.7 in your workbook with common rank 12.5 and then calculate H' as

$$
H' = \frac{H}{1 - \frac{2(2^3 - 1)}{16^3 - 16}} = \frac{H}{.9965686}
$$

Show that the P-value becomes 0.134.

Note: This is the P-value given on page 30 in the Student CD-ROM, which is also adjusted for the presence of ties.

Chapter 15

Logistic Regression

Excel does not possess a built-in logistic regression tool. Logistic regression is a highly specialized topic, best used under the guidance of a statistician. However, it is still possible within Excel to obtain estimates of the parameters in simple logistic regression with little effort using weighted least squares.

15.1 The Logistic Regression Model

Recall from the discussion in Section 2.5 and Chapter 10 that regression refers to fitting models for the mean value of a response as a function of an explanatory variable. For n pairs of observations (x_i, y_i) the simple linear regression model is

$$y_i = \beta_0 + \beta_1 x_i + \varepsilon_i$$

with the errors $\{\varepsilon_i\}$ assumed independent $N(0, \sigma)$ and the regression function

$$\mu_y = \beta_0 + \beta_1 x_i$$

representing the mean of y_i as a function of x.

It is possible to fit models *other* than a straight line. If you examine Fig 2.14, you will observe that Excel can fit

$$\mu_y = a + b \log x \quad \text{(logarithmic)}$$
$$\mu_y = ax^b \quad \text{(power)}$$
$$\mu_y = ae^{bx} \quad \text{(exponential)}$$

as well as polynomial and moving average models. These nonlinear models are fitted using transformations which linearize the regression curve.

In some circumstances, the response variable is discrete, not continuous. An example of a discrete variable might be the number of cases of skin cancer in a metropolitan area. A special case of a discrete variable is a binary response, say $\{0, 1\}$, which leads to the logistic regression model. We restrict attention to binary response in the remainder of this chapter.

Binomial Distributions and Odds

First, we will identify some major differences between regression with binary responses and regression with continuous responses. Assume, as is customary, that the errors have mean 0, which endows the regression curve

$$\mu_y = \beta_0 + \beta_1 x$$

with the usual meaning as the mean response. In the Bernoulli case, y_i can take only two values so that

$$\mu_y = P[y_i = 1] = \beta_0 + \beta_1 x_i = p_i$$

and the variance of the error term is therefore the variance of a Bernoulli random variable, $p_i(1 - p_i)$. Note:

- Errors cannot be normal.
- Errors have nonconstant variance.
- The mean response is constrained to lie within the interval [0,1] since it represents a probability.

Because of these differences, ordinary regression methodology is not appropriate.

The explanatory variables are $(x_1, x_2, \ldots, x_c)$ with n_i observations at the value x_i. The number of "successes" at x_i is denoted by s_i, which is a binomial $\text{Bin}(n_i, p_i)$ random variable on n_i trials and success probability $p_i \equiv p(x_i)$, which is a function of x_i. It is the function $p(x_i)$ that is of interest.

We have dealt with binary responses on two occasions in this text; in Chapter 8 we considered $c = 2$, while in Chapter 9 we dealt with $c \geq 2$. Here we also consider $c \geq 2$ but impose a model (the regression curve) relating all the probabilities p_i as a function of x_i.

A scatterplot of such data shows binary observations having y values of 0 and 1 and the interpretation of the regression curve as somehow passing near the data is lost because of the categorical (binary) nature of the response. However, the interpretation is partially regained if a scatterplot is made of the sample proportions $\hat{p}_i = \frac{s_i}{n_i}$ on the y-axis against their corresponding x_i on the x-axis (as shown in Fig 15.4 on page 44 of the Student CD-ROM).

We may then consider fitting a curve through the $(x_i, \hat{p}_i)$ pairs. In order to facilitate this approach, statisticians have introduced the logit transformation based on the odds. Define the odds ratio

$$\text{ODDS} = \frac{p}{1 - p}$$

and then take the natural logarithm of the odds to define the logit function:

$$\text{logit}(p) = \log(\text{ODDS}) = \log\left(\frac{p}{1 - p}\right)$$

There is a mathematical reason for using ODDS as the *natural* parameter that comes from the factorization of the likelihood function. These are advanced details, beyond the scope of this presentation.

The Statistical Model

The **statistical model for logistic regression** posits a linear form

$$\text{logit}(p) = \beta_0 + \beta_1 x$$

that is equivalent to

$$p \equiv p(x) = \frac{1}{1 + e^{-(\beta_0 + \beta_1 x)}}$$

in terms of the original probability p (Fig 15.4 of the Student CD-ROM).

In the text, estimates for the parameters β_0, β_1 are presented based on the output from a specialized statistics package called SAS. Although Excel does not provide a logistic regression output, we can obtain estimates using the **spreadsheet** features of Excel and **weighted least-squares**.

Weighted Least-Squares

The least-squares criterion minimizes the sum of the squared residuals

$$\sum_{i=1}^{n} e_i^2 = \sum_{i=1}^{n} (y_i - \hat{y}_i)^2$$

where $\{y_i\}$ are the observed values and $\{\hat{y}_i\}$ are the fitted values where $\hat{y}_i = b_0 + b_1 x_i$. When the errors do not have a constant variance, it is more **efficient** (in the sense of producing estimates with smaller variance) to weight the residuals. It is intuitively reasonable to give a residual more weight (for accuracy) if its corresponding variance is smaller.

Let w_i be the weight assigned to an observation at x_i. Then w_i is inversely proportional to the variance σ_i^2 of the error ε_i and the criterion for weighted least-squares is

$$\text{minimize} \quad \sum_{i=1}^{n} w_i (y_i - b_0 - b_1 x_i)^2$$

The solution is obtained (as in Chapter 2) by differentiating with respect to b_0, b_1. We find

$$b_1 = \frac{\sum w_i x_i y_i - \frac{(\sum w_i x_i)(\sum w_i y_i)}{\sum w_i}}{\sum w_i x_i^2 - \sum w_i x_i}$$

$$b_0 = \frac{\sum w_i y_i - b_1 \sum w_i x_i}{\sum w_i}$$

When $w_i \equiv 1$ for all i, these two equations reduce to the equations obtained for finding the ordinary least-squares estimates.

These equations represent the quintessential spreadsheet operations, columns of numbers that are added and whose totals are then algebraically manipulated.

Therefore, they are ideal for spreadsheet logic, and *it is fitting in the final chapter, dealing with a sophisticated statistical model for which Excel does not provide a built-in tool, that we can obtain a solution by setting up our own columns of variables.*

The variable y_i, which is appropriate in this setting, is

$$y_i = \log \frac{\hat{p}_i}{1 - \hat{p}_i}$$

and the weights are approximately $w_i = n_i p_i (1 - p_i)$. Since the $\{p_i\}$ are unknown, we employ instead

$$w_i = n_i \, \hat{p}_i (1 - \hat{p}_i)$$

We organize our workbook with columns for

$$x_i, y_i, w_i, w_i x_i, w_i y_i, w_i x_i^2, w_i y_i, w_i x_i y_i$$

which are then used to determine b_0, b_1 (Fig 15.1).

15.2   Inference for Logistic Regression

Example 15.1. (Examples 15.7–15.8, pages 47–50 in the Student CD-ROM.) An experiment was designed to examine how well the insecticide rotenone kills aphids that feed on the chrysanthemum plant called *macrosiphoniella sanborni*. The explanatory variable is the concentration (in log of mg/l) of the insecticide. At each concentration, approximately 50 insects were exposed. Each insect was either killed or not killed. We summarize the data using the number killed. The response variable for logistic regression is the log odds of the proportion killed. Here are the data:

Concentration (log)	Number of insects	Number killed
0.96	50	6
1.33	48	16
1.63	46	24
2.04	49	42
2.32	50	44

Fit a logistic model

$$p_i = \frac{1}{1 + e^{-(\beta_0 + \beta_1 x_i)}}$$

to the probability that an aphid will be killed as a function of the concentration x_i.

	A	B	C	D	E	F	G	H	I	J	K
1					Logistic Regression by Weighted Least Squares						
2											
3					logit	weight			calculations		
4	x	n	s	p	y	w	w*x	w*y	w*x*y	w*x^2	w*y^2
5	0.96	50	6	0.120	-1.992	5.280	5.069	-10.520	-10.099	4.866	20.960
6	1.33	48	16	0.333	-0.693	10.667	14.187	-7.394	-9.833	18.868	5.125
7	1.63	46	24	0.522	0.087	11.478	18.710	0.999	1.628	30.497	0.087
8	2.04	49	42	0.857	1.792	6.000	12.240	10.751	21.931	24.970	19.262
9	2.32	50	44	0.880	1.992	5.280	12.250	10.520	24.406	28.419	20.960
10				sums		38.705	62.455	4.356	28.033	107.620	66.395
11											
12				=s/n	=LN(p/(1-p))	=n*p*(1-p)	=w*x	=w*y	=w*x*y	=w*x^2	=w*y^2
13											
14		b1=	3.070	=(I10-G10*H10/F10)/(J10-G10^2/F10)							
15		b0=	-4.841	=(H10-B14*G10)/F10							

Figure 15.1: Weighted Least-Squares Estimates for Logistic Regression

Solution. Fig 15.1 shows how we have set up the workbook. Use **Named Ranges** for the labels and variables shown in row 4. The formulas required for calculating the columns are shown in cells D12:K12. The weighted least-squares estimates are in B14:B15, and in C14:C15 we have given the formulas to be entered into B14:B15. (Your workbook will not have C14:C15.)

We find the weighted least-squares estimates to be

$$b_1 = 3.070$$
$$b_0 = -4.841$$

These compare favorably with the SAS output presented in the text

$$b_1 = 3.10$$
$$b_0 = -4.89$$

Exercise. (Examples 15.1–15.4, pages 40–45 in the Student CD-ROM.) In Chapter 8 we presented the results of a survey on binge drinking. The text discusses this example to illustrate logistic regression. Adapt the workbook developed in the previous section to show that the weighted least-squares estimates are

$$b_1 = 0.362$$
$$b_0 = -1.587$$

These are identical to those in your text. Your results should look like those in Fig 15.2.

Exercise. Adapt the workbook shown in Fig 15.1 and the formula = CRITBINOM(n, p, RAND()) used in Section 5.2 to simulate data for a

	A	B	C	D	E	F	G	H	I	J	K
1					**Logistic Regression by Weighted Least Squares**						
2											
3					logit	weight			calculations		
4	x	n	s	p	y	w	w*x	w*y	w*x*y	w*x^2	w*y^2
5	0	9916	1684	0.170	-1.587	1398.012	0.000	-2218.445	0.000	0.000	3520.356
6	1	7180	1630	0.227	-1.225	1259.958	1259.958	-1543.723	-1543.723	1259.958	1891.398
7			sums			2657.970	1259.958	-3762.169	-1543.723	1259.958	5411.753
8											
9				=s/n	=LN(p/(1-p))	=n*p*(1-p)	=w*x	=w*y	=w*x*y	=w*x^2	=w*y^2
10											
11	b1=	0.362	= (I7-G7*H7/F7)/(J7-G7^2/F7)								
12	b0=	-1.587	= (H7-B11*G7)/F7								

Figure 15.2: Binge Drinking

logistic regression. Choose the same values $\{x_i\}$ and $\{n_i\}$ as in Example 15.1. Set the parameters to be

$$\beta_1 = -5.00$$
$$\beta_0 = 3.00$$

and obtain estimates b_1, b_0.

Next, generate repeated samples and construct histograms of the values for b_0 and b_1 and compare them with the true values. What can you conclude about the mean, bias, standard error, and confidence intervals? Construct scatterplots of the simulated values of (b_0, b_1).

Index